MONOGRAPHS ON STATISTICS AND APPLIED PROBABILITY

General Editors

V. Isham, N. Keiding, T. Louis, N. Reid, R. Tibshirani, and H. Tong

1 Stochastic Population Models in Ecology and Epidemiology *M.S. Barlett* (1960)
2 Queues *D.R. Cox and W.L. Smith* (1961)
3 Monte Carlo Methods *J.M. Hammersley and D.C. Handscomb* (1964)
4 The Statistical Analysis of Series of Events *D.R. Cox and P.A.W. Lewis* (1966)
5 Population Genetics *W.J. Ewens* (1969)
6 Probability, Statistics and Time *M.S. Barlett* (1975)
7 Statistical Inference *S.D. Silvey* (1975)
8 The Analysis of Contingency Tables *B.S. Everitt* (1977)
9 Multivariate Analysis in Behavioural Research *A.E. Maxwell* (1977)
10 Stochastic Abundance Models *S. Engen* (1978)
11 Some Basic Theory for Statistical Inference *E.J.G. Pitman* (1979)
12 Point Processes *D.R. Cox and V. Isham* (1980)
13 Identification of Outliers *D.M. Hawkins* (1980)
14 Optimal Design *S.D. Silvey* (1980)
15 Finite Mixture Distributions *B.S. Everitt and D.J. Hand* (1981)
16 Classification *A.D. Gordon* (1981)
17 Distribution-Free Statistical Methods, 2nd edition *J.S. Maritz* (1995)
18 Residuals and Influence in Regression *R.D. Cook and S. Weisberg* (1982)
19 Applications of Queueing Theory, 2nd edition *G.F. Newell* (1982)
20 Risk Theory, 3rd edition *R.E. Beard, T. Pentikäinen and E. Pesonen* (1984)
21 Analysis of Survival Data *D.R. Cox and D. Oakes* (1984)
22 An Introduction to Latent Variable Models *B.S. Everitt* (1984)
23 Bandit Problems *D.A. Berry and B. Fristedt* (1985)
24 Stochastic Modelling and Control *M.H.A. Davis and R. Vinter* (1985)
25 The Statistical Analysis of Composition Data *J. Aitchison* (1986)
26 Density Estimation for Statistics and Data Analysis *B.W. Silverman* (1986)
27 Regression Analysis with Applications *G.B. Wetherill* (1986)
28 Sequential Methods in Statistics, 3rd edition
G.B. Wetherill and K.D. Glazebrook (1986)
29 Tensor Methods in Statistics *P. McCullagh* (1987)
30 Transformation and Weighting in Regression
R.J. Carroll and D. Ruppert (1988)
31 Asymptotic Techniques for Use in Statistics
O.E. Bandorff-Nielsen and D.R. Cox (1989)
32 Analysis of Binary Data, 2nd edition *D.R. Cox and E.J. Snell* (1989)
33 Analysis of Infectious Disease Data *N.G. Becker* (1989)
34 Design and Analysis of Cross-Over Trials *B. Jones and M.G. Kenward* (1989)
35 Empirical Bayes Methods, 2nd edition *J.S. Maritz and T. Lwin* (1989)
36 Symmetric Multivariate and Related Distributions
K.T. Fang, S. Kotz and K.W. Ng (1990)

Monographs on Statistics and Applied Probability 102

Diagnostic Checks in Time Series

37 Generalized Linear Models, 2nd edition *P. McCullagh and J.A. Nelder* (1989)

38 Cyclic and Computer Generated Designs, 2nd edition
J.A. John and E.R. Williams (1995)

39 Analog Estimation Methods in Econometrics *C.F. Manski* (1988)

40 Subset Selection in Regression *A.J. Miller* (1990)

41 Analysis of Repeated Measures *M.J. Crowder and D.J. Hand* (1990)

42 Statistical Reasoning with Imprecise Probabilities *P. Walley* (1991)

43 Generalized Additive Models *T.J. Hastie and R.J. Tibshirani* (1990)

44 Inspection Errors for Attributes in Quality Control
N.L. Johnson, S. Kotz and X. Wu (1991)

45 The Analysis of Contingency Tables, 2nd edition *B.S. Everitt* (1992)

46 The Analysis of Quantal Response Data *B.J.T. Morgan* (1992)

47 Longitudinal Data with Serial Correlation—A State-Space Approach
R.H. Jones (1993)

48 Differential Geometry and Statistics *M.K. Murray and J.W. Rice* (1993)

49 Markov Models and Optimization *M.H.A. Davis* (1993)

50 Networks and Chaos—Statistical and Probabilistic Aspects
O.E. Barndorff-Nielsen, J.L. Jensen and W.S. Kendall (1993)

51 Number-Theoretic Methods in Statistics *K.-T. Fang and Y. Wang* (1994)

52 Inference and Asymptotics *O.E. Barndorff-Nielsen and D.R. Cox* (1994)

53 Practical Risk Theory for Actuaries
C.D. Daykin, T. Pentikäinen and M. Pesonen (1994)

54 Biplots *J.C. Gower and D.J. Hand* (1996)

55 Predictive Inference—An Introduction *S. Geisser* (1993)

56 Model-Free Curve Estimation *M.E. Tarter and M.D. Lock* (1993)

57 An Introduction to the Bootstrap *B. Efron and R.J. Tibshirani* (1993)

58 Nonparametric Regression and Generalized Linear Models
P.J. Green and B.W. Silverman (1994)

59 Multidimensional Scaling *T.F. Cox and M.A.A. Cox* (1994)

60 Kernel Smoothing *M.P. Wand and M.C. Jones* (1995)

61 Statistics for Long Memory Processes *J. Beran* (1995)

62 Nonlinear Models for Repeated Measurement Data
M. Davidian and D.M. Giltinan (1995)

63 Measurement Error in Nonlinear Models
R.J. Carroll, D. Rupert and L.A. Stefanski (1995)

64 Analyzing and Modeling Rank Data *J.J. Marden* (1995)

65 Time Series Models—In Econometrics, Finance and Other Fields
D.R. Cox, D.V. Hinkley and O.E. Barndorff-Nielsen (1996)

66 Local Polynomial Modeling and its Applications *J. Fan and I. Gijbels* (1996)

67 Multivariate Dependencies—Models, Analysis and Interpretation
D.R. Cox and N. Wermuth (1996)

68 Statistical Inference—Based on the Likelihood *A. Azzalini* (1996)

69 Bayes and Empirical Bayes Methods for Data Analysis
B.P. Carlin and T.A Louis (1996)

70 Hidden Markov and Other Models for Discrete-Valued Time Series
I.L. Macdonald and W. Zucchini (1997)

71 Statistical Evidence—A Likelihood Paradigm *R. Royall* (1997)

72 Analysis of Incomplete Multivariate Data *J.L. Schafer* (1997)

73 Multivariate Models and Dependence Concepts *H. Joe* (1997)

74 Theory of Sample Surveys *M.E. Thompson* (1997)

75 Retrial Queues *G. Falin and J.G.C. Templeton* (1997)

76 Theory of Dispersion Models *B. Jørgensen* (1997)

77 Mixed Poisson Processes *J. Grandell* (1997)

78 Variance Components Estimation—Mixed Models, Methodologies and Applications *P.S.R.S. Rao* (1997)

79 Bayesian Methods for Finite Population Sampling *G. Meeden and M. Ghosh* (1997)

80 Stochastic Geometry—Likelihood and computation *O.E. Barndorff-Nielsen, W.S. Kendall and M.N.M. van Lieshout* (1998)

81 Computer-Assisted Analysis of Mixtures and Applications—Meta-analysis, Disease Mapping and Others *D. Böhning* (1999)

82 Classification, 2nd edition *A.D. Gordon* (1999)

83 Semimartingales and their Statistical Inference *B.L.S. Prakasa Rao* (1999)

84 Statistical Aspects of BSE and vCJD—Models for Epidemics *C.A. Donnelly and N.M. Ferguson* (1999)

85 Set-Indexed Martingales *G. Ivanoff and E. Merzbach* (2000)

86 The Theory of the Design of Experiments *D.R. Cox and N. Reid* (2000)

87 Complex Stochastic Systems *O.E. Barndorff-Nielsen, D.R. Cox and C. Klüppelberg* (2001)

88 Multidimensional Scaling, 2nd edition *T.F. Cox and M.A.A. Cox* (2001)

89 Algebraic Statistics—Computational Commutative Algebra in Statistics *G. Pistone, E. Riccomagno and H.P. Wynn* (2001)

90 Analysis of Time Series Structure—SSA and Related Techniques *N. Golyandina, V. Nekrutkin and A.A. Zhigljavsky* (2001)

91 Subjective Probability Models for Lifetimes *Fabio Spizzichino* (2001)

92 Empirical Likelihood *Art B. Owen* (2001)

93 Statistics in the 21st Century *Adrian E. Raftery, Martin A. Tanner, and Martin T. Wells* (2001)

94 Accelerated Life Models: Modeling and Statistical Analysis *Vilijandas Bagdonavicius and Mikhail Nikulin* (2001)

95 Subset Selection in Regression, Second Edition *Alan Miller* (2002)

96 Topics in Modelling of Clustered Data *Marc Aerts, Helena Geys, Geert Molenberghs, and Louise M. Ryan* (2002)

97 Components of Variance *D.R. Cox and P.J. Solomon* (2002)

98 Design and Analysis of Cross-Over Trials, 2nd Edition *Byron Jones and Michael G. Kenward* (2003)

99 Extreme Values in Finance, Telecommunications, and the Environment *Bärbel Finkenstädt and Holger Rootzén* (2003)

100 Statistical Inference and Simulation for Spatial Point Processes *Jesper Møller and Rasmus Plenge Waagepetersen* (2004)

101 Hierarchical Modeling and Analysis for Spatial Data *Sudipto Banerjee, Bradley P. Carlin, and Alan E. Gelfand* (2004)

102 Diagnostic Checks in Time Series *Wai Keung Li* (2004)

Monographs on Statistics and Applied Probability 102

Diagnostic Checks in Time Series

Wai Keung Li

CHAPMAN & HALL/CRC

A CRC Press Company

Boca Raton London New York Washington, D.C.

Library of Congress Cataloging-in-Publication Data

Li, Wai Keung
Diagnostic checks in time series / Wai Keung Li.
p. cm. -- (Monographs on statistics and applied probability ; 102)
Includes bibliographical references and index.
ISBN 1-58488-337-5 (alk. paper)
1. Time-series analysis. I. Title. II. Series.

QA280.L5 2004
519.5′.5—dc22 2003063471

International Standard Book Number 1-58488-337-5
Library of Congress Card Number 2003063471
Printed in the United States of America 1 2 3 4 5 6 7 8 9 0
Printed on acid-free paper

To my family,

my parents and the memory of my grandparents

Contents

Preface

This book is about diagnostic checking for time series models over discrete time. There are many texts and monographs on time series modeling but almost none of them has diagnostic checking as the major focus. Hence, it is hoped that the present book will fill an important gap in the literature. This book focuses mainly on diagnostic checks for stationary time series. Therefore, topics such as unit root and co-integration tests and diagnostic checks for spatial time series have not been included. However, unit root and co-integration tests have been well covered by many authors. Indeed, even though we have only stationary time series in mind, the literature for diagnostic checks is very extensive and a further narrowing of the focus is necessary. As a result, we only mention outlier detection in passing because this topic has a large literature. Nevertheless, this book covers many different time series models including the univariate and multivariate autoregressive moving-average (ARMA) models, the threshold type time series models, the bilinear models, exponential autoregressive models, models with conditional autoregressive heteroscedasticity (ARCH), long memory or fractionally integrated ARMA models, conditional non-Gaussian models and the autoregressive conditional duration models. A major theme of the book is the portmanteau goodness-of-fit test which appears in slightly different forms in almost all situations. Much criticism has been levelled at the possible low power of this type of pure significance test. However, it remains a useful and important diagnostic tool for time series models for the following reasons. First, like the classical sample mean it is easy to understand conceptually and falls in line with the traditional approach to data analysis. In most situations it is also fairly easy to compute. To me, it provides a challenge in the modeling of time series. Second, as the present book demonstrates, such a test exists for nearly all situations. Like Pearson's classic goodness-of-fit tests it can be adapted or constructed for most situations. Of course, the score test enjoys a similar status and is also discussed extensively in this book.

This book also reflects my personal learning process. Through the years I have learned a lot from various people. I am greatly indebted to my mentor, Professor A.I. McLeod, of the University of Western Ontario, for introducing me to the original portmanteau test in the ARMA case and many other interesting aspects of time series. Through the years, it has become clear to me that his method of deriving the test is, in fact, as powerful as the test's versatility. Without his initial guidance this book would not have been possible. Unlike many other monographs the current book is not a consequence of a lecture course in time series. However, I trust that it can also be used in this way as an introduction to various time series models building on a first course on ARMA models. I have approached the topics in the book with the eyes of a model builder and not as a mathematical statistician. I hope this approach will make the book more accessible to practitioners. Because of the time constraint I have not been able to provide more examples and computer program. Fortunately, most contemporary computer software has readily available procedures to fit most of the models discussed. The recent books by R. Tsay and N.H. Chan also contain useful programs for fitting many of the models and they help compensate for some of the deficiencies of the book.

I would like to thank many people without whose help and encouragement this book would not have been possible. First, I would like to thank Professor A.I. McLeod for his mentorship during my days as a research student in 1978 and for his advice through the years. I would also like to thank Professor Gene Denzel, York University, Canada for teaching me the first course in statistics and offering me financial support when I was an undergraduate student at York. Second, I would like to express my sincere thanks to Professor Howell Tong and my Head of Department Dr. Kai Ng, for their encouragement and support; Terence Chong, Tom Fong, Andy Kwan, Ian Lauder, Heung Wong, Philip Yu, and three reviewers for reading the manuscript and correcting many of my foolish mistakes; Ms. Ada Lai for her expert and skillful typing of the manuscript; Wilson Li for his technical assistance; Peter Brockwell, K.S. Chan, N.H. Chan, W.S. Chan, C.W.J. Granger, Y.V. Hui, Anthony Kuk, K. Lam, Tony Lawrance, Johannes Ledolter, Shiqing Ling, T.K. Mak, Michael McAleer, Peter Robinson, George Tiao, Dag Tjostheim, R. Tsay, Yuk Tse, H. Yang, Kam Yuen, Y. Xia and my research students at the University of Hong Kong from whom I have learned a lot. Third, I would like to express my gratitude to the editorial staff at Chapman & Hall/CRC Press for their help and assistance in making the book possible. Fourth, I would like to thank all the publishers for permission to use materials from papers that have appeared in their journals. I am also grateful to the Hong Kong Research Grants Council and the Committee on Research and Conference Grants of the University of Hong Kong for financial support for my research related to the present work.

Last but not least, I would like to thank my wife Julia and my son Ka Shun for their love, understanding and patience while I was preparing the manuscript for this book.

W.K. Li

The University of Hong Kong

CHAPTER 1

Introduction

One of the major tasks of a statistician is to come up with a probability model that can adequately describe his data. Hence, the often asked question, "which model describes the data best?" or equivalently, "which model provides the best fit to the data?" In the days of Karl Pearson where the emphasis was usually on whether the data are from a certain distribution family this question translates into testing the hypothesis that the common distribution $F(\cdot)$, of an independent identically distributed sample $X_1, \ldots, X_n$ is equal to a family of distribution indexed by, say, a parameter θ. That is, we test the null hypothesis $H_0 : F(\cdot) = G(\cdot|\theta)$. This gives rise to Pearson's 1900 paper on the classical chi-squared goodness-of-fit test. Since then there evolves a huge literature on goodness-of-fit tests. As Moore (1978) pointed out "chi-square tests remain among the most common tests of fit, largely because of the flexibility of Pearson's idea. If, for example, observations $\boldsymbol{X}_j$ and the cells E_i are multidimensional, the distribution of the cell frequencies N_i and the form and theory of the Pearson chi-square statistic are unchanged." Modern statistics have developed many more tools than the chi-square tests in order to answer the question, "which model(s) describes the data more adequately?" Atkinson (1986) suggested that in regression "diagnostics is the name given to a collection of techniques for detecting disagreement between a regression model and the data to which it is fitted." The same can be said about time series analysis. The same classical question "which model best describes the data?" is asked by both theorists and practitioners. The so-called Box-Jenkins approach to time series modeling (Box and Jenkins, 1970; 1976) reflects both the influences of the classical goodness-of-fit and diagnostic approaches. Their approach can be described by the following flowchart (Fig. 1.1):

In the first stage a preliminary autoregressive moving average (ARMA) model is suggested based on information on the sample path, sample moments: autocorrelations and partial autocorrelations. Usually differencing will be performed to transform the data into stationarity. The degree of differencing can be determined graphically as was advocated

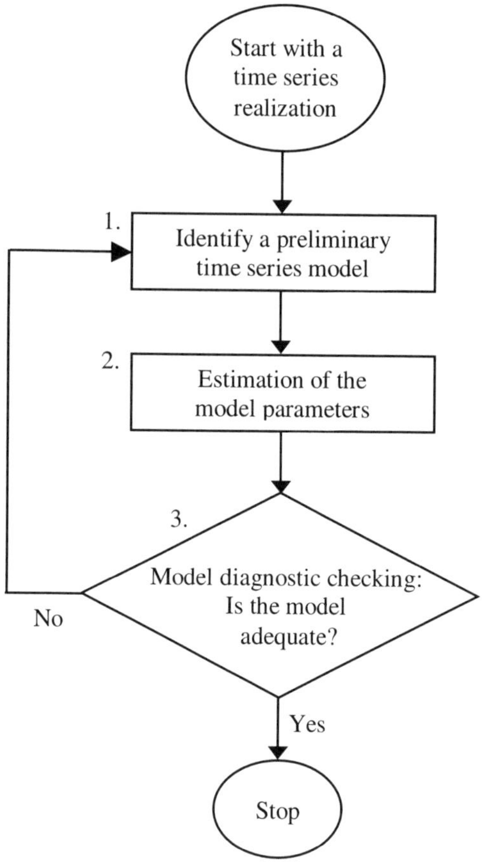

Figure 1.1 *A 3-stage approach to time series modeling*

by Box and Jenkins (1976). However, formal unit root tests can be performed these days routinely to determine the degrees of differencing. See for example Fuller (1996). There is a huge literature on unit root testing and the topic is well treated in many text books and monographs. This book assumes that such a transformation to stationarity has been performed and concentrates mainly on stationary time series. In the second stage, the estimation of stationary ARMA models can be done efficiently by many software routines. At present approximate or exact maximum likelihood procedures are often used for estimation once the autoregressive and moving average orders are specified. For well-specified models, estimation is often not a problem. For pure autoregressive models there are at least two more choices in terms of methods of estimation: the least squares procedure and the Yule-Walker equations. For series of the order of 200 observations, the differences between these methods are small unless the stationarity or invertibility criteria are violated. The third stage in the Box-Jenkins approach is called model diagnostic checking which involves techniques like overfitting, residual plots, and more importantly, checkings that the residuals are approximately uncorrelated. This makes good modeling sense since in the time series analysis a good model should be able to describe the dependence structure of the data adequately, and one important measurement of dependence is via the autocorrelation function. In other words, a good time series model should be able to produce residuals that are approximately uncorrelated, that is, residuals that are approximately white noise. Note that as in the classical regression case complete independence among the residuals is impossible because of the estimation process. However, the autocorrelations of the residuals should be close to being uncorrelated after taking into account the effect of estimation. As shown in the seminal paper by Box and Pierce (1970), the asymptotic distribution of the residual autocorrelations plays a central role in checking out this feature. From the asymptotic distribution of the residual autocorrelations we can also derive tests for the individual residual autocorrelations and overall tests for an entire group of residual autocorrelations assuming that the model is adequate. These overall tests are often called portmanteau tests, reflecting perhaps that they are in the tradition of the classical chi-square tests of Pearson. The latter group of tests has been called omnibus tests by M.S. Bartlett (Cox, 2002). Some of the diagnostic tests introduced in this book are derived under specific type of departures (alternatives) from the null hypothesis and would therefore be more powerful if such departures are in fact true. Nevertheless, portmanteau tests remain useful as an overall benchmark assuming the same kind of role as the classical chi-square tests. It can also be seen that like the classical chi-square tests, portmanteau tests or their variants can be derived under a variety of situations. Portmanteau tests and the residual autocorrelations are easy to compute and the rationale of using them is easy to understand. These considerations enhance their usefulness in applications. Of course, many portmanteau tests can also be derived as tests against specific alternatives.

This book assumes that the reader has already taken a course in elementary time series analysis. A good course in time series based on the books by Cryer (1986), Abraham and Ledolter (1983, Ch.5–8), or Wei (1990, Ch.1–10) should be able to provide sufficient background. Brockwell and Davis (1996) also provides a good and rigorous beginning. One feature of good about Brockwell and Davis is that it comes together with a good software package for ARMA modeling which is user friendly and has good diagnostic checking features. Although reading the present book requires some background in time series to begin with, our orientation and motivation are more on the applied side.

Model diagnostic checkings are often used together with model selection criteria such as the Akaike information criterion (AIC) and the Bayesian information criterion (BIC). These two approaches actually complement each other. Model diagnostic checks can often suggest directions to improve the existing model while information criteria can be used in a more or less "automatic" way within the same family of models. There is already a comprehensive treatment on model selection by McQuarrie and Tsai (1998) and hence the present book will concentrate on the other side of the story, diagnostic checks for time series models. Many time series models are introduced along with the respective diagnostic checking procedures in the following chapters. Through the exposition on diagnostic checking methods, it is hoped that the practitioner should be able to grasp the relative merits of these models and how these different models can be estimated. Hence, answering the question "Which model describes the data best?" The arrangement of the book is as follows.

Chapter 2 considers diagnostic tests for univariate ARMA type models. The relationship between the portmanteau test and the Lagrange multiplier test is also discussed. Extension of the portmanteau test to periodic autoregression is included as well as a new test due to Peña and Rodriǵuez (2002). Chapter 3 considers the multivariate ARMA case and tests for the so-called Granger causality. In Chapter 4 the robustified versions of the residual autocorrelations and portmanteau tests are considered. Chapter 5 considers some popular nonlinear time series models. Diagnostic tests for the possible presence of nonlinearity and goodness-of-fit tests for nonlinear models are discussed. The difficult problem of choosing between two different families of nonlinear models is also discussed briefly. Chapter 6 considers diagnostic checks for the presence of conditional heteroscedasticity which is often modeled by the so-called autoregressive conditional heteroscedastic (ARCH) models and also goodness-of-fit tests for both univariate and multivariate ARCH type models. In Chapter 7 the long memory or fractionally differenced ARMA (FARIMA) models are considered. Finally, in Chapter 8 a variety of non-Gaussian models are considered including conditional models based on the generalized linear model and the autoregressive duration models. Finally, a recently proposed transformation that seems to be able to improve the power performance of diagnostic tests is also introduced.

CHAPTER 2

Diagnostic checks for univariate linear models

2.1 Introduction

One of the most successful statistical models ever developed for time series data is the autoregressive moving average (ARMA) model. Its popularity began in the early 1970s partly due to the book by Box and Jenkins and partly due to advancement in computing power which allows the likelihood function of the ARMA models to be evaluated efficiently. Now most commercial statistical softwares are capable of ARMA time series modeling.

In this chapter we will assume that the time series $\{X_t\}$ satisfies the ARMA (p, q) model

$$X_t - \phi_1 X_{t-1} - \cdots - \phi_p X_{t-p} = \theta_0 + a_t - \theta_1 a_{t-1} - \cdots - \theta_q \, a_{t-q} \quad (2.1)$$

where a_t is white noise with mean 0, variance σ^2, and finite fourth order moment. It is further assumed that $\{X_t\}$ is stationary, invertible, and identifiable. Denote by B the backward shift operator, $B\,X_t = X_{t-1}$. The necessary and sufficient condition for second order stationarity is that the polynominal $\phi(B) = 1 - \phi_1 B - \cdots - \phi_p B^p$ has all roots outside the unit circle. Similarly, the necessary and sufficient condition for invertibility is that all roots of the polynomial $\theta(B) = 1 - \theta_1 B - \cdots - \theta_q B^q$ lie outside the unit circle. For identifiability we require that $\phi(B)$ and $\theta(B)$ have no common roots.

It is easily seen from (2.1) that $\theta_0 = (1 - \phi_1 \cdots - \phi_p)\mu$ where $\mu = E(X_t)$. In most cases, there is no loss of generality to assume that $\mu = 0$. In terms of the backshift operator B (2.1) can be rewritten as

$$\phi(B) X_t = \theta(B) a_t \; . \quad (2.2)$$

In most applications a_t is assumed to be Gaussian and given $X_1, \ldots, X_n$, asymptotically efficient estimation of the parameters ϕ_i, $i = 1, \ldots, p$; θ_j, $j = 1, \ldots, q$ can be achieved by maximizing the conditional log-likelihood

$$l = \text{constant} - \frac{1}{2} n \ln \sigma^2 - \frac{1}{2\sigma^2} \sum_{t=1}^{n} a_t^2 \,. \tag{2.3a}$$

Now (2.3a) can be maximized with respect to σ^2, $\phi_1, \ldots, \phi_p$, $\theta_1, \ldots, \theta_q$ by assuming that $X_1, \ldots, X_p$, are fixed and $a_t = 0$ for $t \leq p$. If (2.3a) is maximized first with respect to σ^2 it can be seen that the maximum likelihood estimator of σ^2 is $\hat{\sigma}^2 = \sum_{t=p+1}^{n} a_t^2/n$. Substituting $\hat{\sigma}^2$ into (2.3a) it can be seen that apart from a constant

$$l_{(\max)} = -\frac{n}{2} \ln \sum_{p+1}^{n} a_t^2 \,.$$

Hence, maximizing the concentrated log-likelihood $l_{(\max)}$ with respect to $\phi_1, \ldots, \phi_p$, $\theta_1, \ldots, \theta_q$ is equivalent to minimizing the conditional sum of squares

$$S^2 = \sum_{t=p+1}^{n} a_t^2 \tag{2.3b}$$

where $a_t = X_t - \phi_1 X_{t-1} \cdots - \phi_p X_{t-p} + \theta_1 a_{t-1} + \cdots + \theta_q a_{t-q}$, with $a_t = 0$, if $t \leq p$. Hence, estimates of ϕ_i, θ_j obtained by minimizing S^2 are asymptotically efficient under the Gaussian assumption. If the length of realization is short exact likelihood estimation or the unconditional (backcasting) least squares procedure is recommended. See Box and Jenkins (1976), Brockwell and Davis (1991), and Box, Jenkins, and Reinsel (1994) for more details on estimation. See also McLeod (1977). The residuals resulting from the fitted model are denoted by $\hat{a}_t$.

In the Box-Jenkins approach to ARMA time series modeling it is important to perform diagnostic checking on the residuals of the fitted model. This usually consists of a group of tests including tests for normality using the residuals $\hat{a}_t$. In this connection, the residual skewness (K_3) and kurtosis (K_4) are often employed. These are defined by

$$K_3 = \left(\frac{1}{n} \sum_{t=1}^{n} \hat{a}_t^3 \right) \Big/ \left(\frac{1}{n} \sum_{t=1}^{n} \hat{a}_t^2 \right)^{3/2}$$

and

$$K_4 = \left(\frac{1}{n} \sum_{t=1}^{n} \hat{a}_t^4 \right) \Big/ \left(\frac{1}{n} \sum_{t=1}^{n} \hat{a}_t^2 \right)^{2} - 3 \,.$$

Under the assumption of normality and if the model is correct, K_3 has an asymptotic normal distribution with mean 0 and variance $6/n$ and K_4 has an asymptotic normal distribution with mean 0 and variance $24/n$. Pierce (1985) showed that the asymptotic results are good for first and second order autoregressive processes with sample size as small as 20. The treatise by Hipel and McLeod (1994, Ch.7) contains more discussions on these statistics. However, these features are less important in time series and the most frequently employed test statistic is the residual autocorrelation function

$$\hat{r}_k = \sum_{t=k+1}^{n} \hat{a}_t \, \hat{a}_{t-k} \Bigg/ \sum_{t=1}^{n} \hat{a}_t^2 \,, \tag{2.4}$$

$k = 1, \ldots, m$.

If the model is adequate and $n \gg m$, it is expected that $\hat{r}_1 \cong \hat{r}_2 \cong \cdots \hat{r}_m \cong 0$. Tests of adequacy of the model can therefore be based on the magnitudes of $\hat{r}_k$, the rationale being that a "good" model should produce residuals that are uncorrelated approximately, at least. Clearly, formal tests of goodness-of-fit have to be based on the sampling distribution of $\hat{\boldsymbol{r}} = (\hat{r}_1, \ldots, \hat{r}_m)^{\mathrm{T}}$, where the superscript "T" refers to the transpose of a vector or a matrix.

2.2 The asymptotic distribution of the residual autocorrelation distribution

The asymptotic distribution of the residual autocorrelation $\hat{\boldsymbol{r}}$ from univariate ARMA models was first derived by Box and Pierce (1970). (As noted by Hosking (1978), Walker (1952) was the first to obtain the distribution under the autoregressive model.) Their result is further refined and extended to the multiplicative seasonal ARMA models by McLeod (1978). It is instructive to consider McLeod's result in this chapter. It can be shown that the large sample Fisher information matrix $\mathbf{I}$ for any asymptotically efficient estimator $\hat{\boldsymbol{\beta}}$ of $\boldsymbol{\beta} = (\phi_1, \ldots, \phi_p, \theta_1, \ldots, \theta_q)^{\mathrm{T}}$ can be written (assuming $\sigma^2 = 1$),

$$\mathbf{I} = \left(\begin{array}{c|c} \gamma_{vv}(i-j) & \gamma_{vu}(i-j) \\ \hline \gamma_{uv}(i-j) & \gamma_{uu}(i-j) \end{array}\right)_{(p+q)\times(p+q)} \tag{2.5}$$

where γ_{vv}, γ_{vu}, γ_{uv}, and γ_{uu} are the theoretical autocovariance and cross-covariance of the process u_t and v_t defined by

$$\phi(B)v_t = -a_t \tag{2.6}$$

and

$$\theta(B)u_t = a_t \,.$$

That is, $\gamma_{uu}(k) = E(u_t u_{t+k})$, $\gamma_{vv}(k) = E(v_t v_{t+k})$, and $\gamma_{uv}(k) = E(u_t v_{t+k})$ respectively where $E(\cdot)$ denotes the expectation operator. In other words, the upper $p \times p$ block of $\mathbf{I}$ corresponds to $\boldsymbol{\phi} = (\phi_1, \ldots, \phi_p)^{\mathrm{T}}$ and the lower $q \times q$ block of $\mathbf{I}$ corresponds to $\boldsymbol{\theta} = (\theta, \ldots, \theta_q)^{\mathrm{T}}$. Let $\boldsymbol{r}$ be the counterpart of $\hat{\boldsymbol{r}}$ with $\hat{a}_t$ replaced by a_t. That is $\boldsymbol{r} = (r_1, r_2, \ldots, r_m)^{\mathrm{T}}$ where $r_k = \sum_{t=k+1}^{n} a_t \, a_{t-k} \big/ \sum_{t=1}^{n} a_t^2$. It is well known that the large sample distribution of $\sqrt{n} \cdot \boldsymbol{r}$ is multivariate normal with mean $\mathbf{0}_{m\times 1}$ and covariance matrix $\mathbf{1}_m$, the $m \times m$ identity matrix. Let the power series expansions of $1/\phi(B) = \phi^{-1}(B) = \sum_{l=0}^{\infty} \phi_l' B^l$ and the power series expansions of $1/\theta(B) = \theta^{-1}(B) = \sum_{l=0}^{\infty} \theta_l' B^l$ with $\phi_l' = \theta_l' = 0$ if $l < 0$. Define the $m \times (p+q)$ matrix

$$\mathbf{X} = \left(-\,\phi_{i-j}' \mid \theta_{i-j}' \right) , \tag{2.7}$$

where ϕ_{i-j}' is the (i,j)th element of X and θ_{i-j}' is the $(p+i,j)$th element of X, $j = 1, \ldots, m$. We then have the following theorem.

Theorem 2.1 *(McLeod, 1978) The large sample distribution of $\hat{\boldsymbol{r}}$ is normal with mean $\mathbf{0}_{m\times 1}$ and covariance matrix*

$$\mathbf{C} = \mathrm{var}(\hat{\boldsymbol{r}}) = \frac{1}{n}\left(\mathbf{1}_m - \mathbf{X}\,\mathbf{I}^{-1}\,\mathbf{X}^{\mathrm{T}}\right) . \tag{2.8}$$

Theorem 2.1 follows from the following lemma in McLeod (1978).

Lemma 2.1 *The joint asymptotic distribution of $\sqrt{n}(\hat{\boldsymbol{\beta}} - \boldsymbol{\beta}, \boldsymbol{r})$ is normal with mean $\mathbf{0}$ and covariance matrix*

$$\begin{array}{cc} \left(\begin{array}{c|c} \mathbf{I}^{-1} & -\,\mathbf{I}^{-1}\mathbf{X}^{\mathrm{T}} \\ \hline -\mathbf{X}\mathbf{I}^{-1} & \mathbf{1} \end{array}\right) & \begin{array}{c} p+q \\ m \end{array} \\ \begin{array}{cc} \quad p+q & \qquad m \end{array} & \end{array}$$

where $\mathbf{I}, \mathbf{X}$ and $\mathbf{1}$ are as in Theorem 2.1.

The lemma can be proven by showing that

$$\hat{\boldsymbol{\beta}} - \boldsymbol{\beta} = \mathbf{I}^{-1}\mathbf{S}_{\mathbf{C}} + O_p\big(\frac{1}{n}\big) \tag{2.9}$$

where $\mathbf{S}_{\mathbf{C}}$ is a $p+q$ vector with i^{th} element $-\sum a_t v_{t-i}/n$, $1 \le i \le p$, $-\sum a_t u_{t-p-i}/n$, if $p+1 \le i \le p+q$; and that

$$\hat{\boldsymbol{r}} = \boldsymbol{r} + \mathbf{X}(\hat{\boldsymbol{\beta}} - \boldsymbol{\beta}) + O_p\big(\frac{1}{n}\big) . \tag{2.10}$$

By standard techniques it can then be shown that the asymptotic covariance matrix of $\sqrt{n}(\hat{\boldsymbol{\beta}} - \boldsymbol{\beta})$ and $\sqrt{(n)}\boldsymbol{r}$ is $-\mathbf{I}^{-1}\mathbf{X}^{\mathrm{T}}$.

Note that in Theorem 2.1 for m large enough such that $\phi_l' \cong \mathbf{0}$ and $\theta_l' \cong 0$ for $l > m$, $\mathbf{X}^{\mathrm{T}}\mathbf{X} \cong \mathbf{I}$ and therefore $n \cdot \mathrm{var}(\hat{\boldsymbol{r}})$ is approximately idempotent of rank $m-p-q$. Hence, by a characterization of the multivariate normal distribution (Rao, 1973) the portmanteau or Box-Pierce statistic

$$Q_m = n \cdot (\hat{\boldsymbol{r}}^{\mathrm{T}}\hat{\boldsymbol{r}}) = n \sum_{k=1}^{m} \hat{r}_k^2 \tag{2.11}$$

is asymptotically chi-squared distributed with $m-p-q$ degrees of freedom if the fitted model is adequate, i.e., the fitted model provides approximately uncorrelated residuals. In other words, the model is considered to have fitted the data well if all the residual autocorrelations can be regarded as insignificantly different from zero. Note also that from (2.8) and if X_t is an autoregressive process of order one the asymptotic variance of $\hat{r}_1$ is given by ϕ_1^2/n which, as was observed by Box and Pierce (1970), can be substantially smaller than $1/n$, so that using $1/\sqrt{n}$ as the standard error for $\hat{r}_1$ could result in a conservative evaluation of the adequacy of the model. Similarly for $\{X_t\}$ satisfying a moving average process of order one, the large sample variance of $\hat{r}_1$ is θ_1^2/n. In practice, we replace ϕ_1 or θ_1 by the respective estimators $\hat{\phi}_1$ or $\hat{\theta}_1$. In general, the Fisher information matrix $\mathbf{I}$ can be computed by an algorithm due to McLeod (1975). See also Ansley (1980).

As is remarked in McLeod (1978), if any subset of ϕ_j, $1 \le j \le p$, or θ_j, $1 \le j \le q$, are constrained to zero, then the asymptotic covariance of $\hat{\boldsymbol{r}}$ can be obtained from

$$\left(\mathbf{1}_m - \mathbf{X}_0 \mathbf{I}_0^{-1} \mathbf{X}_0^{\mathrm{T}}\right)/n$$

where $\mathbf{I}_0$ is obtained from $\mathbf{I}$ by deleting the rows and columns corresponding to the constrained parameters and $\mathbf{X}_0$ is obtained from $\mathbf{X}$ by deleting the corresponding columns. This result also implies that Q_m is asymptotically distributed as χ^2 with $m-p_0-q_0$ degrees of freedom if the model is adequate, where p_0 and q_0 are respectively the number of estimated autoregressive and moving average parameters.

Note also that the degrees of freedom of Q_m remain unchanged whether one estimates the mean of $\{X_t\}$ or not. This follows directly from the result of Pierce (1972) on diagnostic checking in transfer function noise models. In some textbooks, the degrees of freedom of Q_m are set equal to $m-p-q-1$ when the mean or the intercept θ_0 is estimated while the degrees of freedom of Q_m are set equal to $m-p-q$ when X_t is centered by its sample mean. The two procedures are actually asymptotically equivalent and the degrees of freedom should be $m-p-q$ whether or not a mean is subtracted from X_t.

Shin and Lee (1996) considered an extension of Theorem 2.1 to nonstationary autoregressive models. In particular they showed that the limiting distribution of the residual autocorrelations are the same as the limiting distribution when parameters are estimated with all roots on the unit circle known. Runde (1997) considered the distribution of Q_m for series with infinite variance. In this case Q_m is no longer χ^2 distributed asymptotically but tends to a complicated limiting distribution for X_t in the domain of attraction of a stable law with characteristic exponent

α, $1 < \alpha < 2$. In particular, it was shown that $\sqrt{n}\hat{r}_k \to 0$ as $n \to \infty$ so that $Q_m \to 0$ as $n \to \infty$. This suggests that a different norming constant for $\hat{r}_k$ other than $\sqrt{n}$ should be considered. The treatment is beyond the scope of this monograph.

2.3 Modifications of the portmanteau statistic

In the previous section it has been shown that the statistic Q_m is asymptotically chi-squared distributed with $m-p-q$ degrees of freedom if the model is adequate and $m \gg 0$. Chatfield (1976), in the discussion of the paper by Prothero and Wallis, questioned the validity of the distribution for a finite n. Davies, Triggs, and Newbold (1977) further demonstrated Q_m could be too conservative in practice even for a moderate n. Ljung and Box (1978) and Prothero and Wallis (1976) advocated the use of the modified statistic

$$\tilde{Q}_m = n(n+2)\sum_{k=1}^{m} \hat{r}_k^2 \,/(n-k) \ . \tag{2.12}$$

The statistic $\tilde{Q}_m$ has a finite sample distribution that is much closer to that of χ^2_{m-p-q}. This modification of the Q_m statistic has since been adopted by many practitioners and is often referred to as the Ljung-Box statistic or the Ljung-Box-Pierce statistic. The motivation for the modification is from the fact that $\text{var}(\hat{r}_k) \cong (n-k)/\{n(n+2)\}$. That is, $\tilde{Q}_m$ is obtained by adjusting essentially each of the $\hat{r}_k$ in Q_m by its asymptotic variance. However, this modification is not without criticism. Davies, Triggs, and Newbold (1977) showed that the variance of $\tilde{Q}_m$ could be substantially larger than that of a chi-squared distribution with $m-p-q$ degrees of freedom, viz., $2(m-p-q)$.

Li and McLeod (1981) suggested an alternative modification by observing that

$$n \cdot \sum_{k=1}^{m} E(\hat{r}_k^2) \cong m\left(1 - \frac{m+1}{2n}\right) \ .$$

The second term could be quite substantial if $2n$ is not much greater than $m(m+1)$. Therefore, Li and McLeod (1981) recommended the modification

$$Q_m^* = Q_m + \frac{m(m+1)}{2n} \ . \tag{2.13}$$

One advantage of Q_m^* is that unlike $\tilde{Q}_m$ it moves the finite sample distribution of Q_m much closer to its asymptotic mean without inflating its variance. Q_m^* is also very easy to apply and program although it is less popular than $\tilde{Q}_m$. Kheoh and McLeod (1992) demonstrated via simulation the advantage of Q_m^*. They compared empirically the significance level, the mean, and the variances of $\tilde{Q}_m$ and Q_m^*. Q_m^* has, in general, a variance that is closer to the variance of the asymptotic chi-square distribution whereas $\tilde{Q}_m$ is more sensitive with significance levels somewhat larger than the nominal levels when n is large. In contrast Q_m^* is slightly conservative. However, the powers of the two tests are almost identical, with the power of $\tilde{Q}_m$ slightly higher. They also suggest that in practice a conservative test is preferred to one that is sensitive. This is particularly the case when their power is comparable. This modification has been incorporated into the McLeod-Hipel time series package.

Example 2.1 *The model $X_t = (1 - 0.4B)a_t$ was fitted to a series of $n = 80$ observations using the exact maximum likelihood procedure. The first 10 residual autocorrelations are listed below.*

k	1	2	3	4	5	6	7	8	9	10
$\hat{r}_k$	.4	.15	.07	.06	.09	.03	.05	.06	.5	.01

The portmanteau statistic Q_m using $m = 10$ is given by

$$\begin{aligned} Q_m &= 80\left(.4^2 + .15^2 + \cdots .05^2 + .01^2\right) \\ &= 16.696 \ . \end{aligned}$$

The upper 5% critical value from the chi-squared distribution with 9 degrees of freedom is 16.92. Therefore, based on Q_m the model is marginally adequate. However, using $\hat{\theta}$ the asymptotic variance of $\hat{r}_1$ is equal to $0.4^2/80$ which gives an asymptotic standard error of 0.045, suggesting that $\hat{r}_1$ is significantly different from zero. Using Li and McLeod (1981) the statistic Q_m is easily adjusted to be

$$\begin{aligned} Q_m^* &= 16.696 + \frac{10(11)}{160} \\ &= 17.384 \end{aligned}$$

whereas

$$\tilde{Q}_m = 80(82)\left(.4^2/79 + .15^2/78 + \cdots + .01^2/70\right) = 17.488 \ .$$

Both Q_m^* and $\tilde{Q}_m$ are significant at the 5% significance level. The adjustment $\tilde{Q}_m$ is somewhat more involved than Q_m^*.

Ljung (1986) considered modifications based on the eigenvalues of the covariance matrix $\mathbf{C}$ of $\hat{\boldsymbol{r}}$ in (2.8). Using a theorem on quadratic forms it was shown that

$$Q_m \sim \sum_{i=1}^{m} \lambda_i \chi_{1,i}^2 \ ,$$

where λ_i are the eigenvalues of $n\,C$, $\chi^2_{1,i}$ are independent χ^2_1 random variables, and '$\sim$' means that the variable on the right-hand side has the same distribution as the one on the left. For a first order AR(1) process with parameter ϕ,

$$Q_m \sim \chi^2_{m-1} + \phi^{2m}\chi^2_1 \, .$$

Ljung suggested a modification of the Q_m distribution using an $a\,\chi^2_b$ distribution with $a = \sum \lambda_i^2 / \sum \lambda_i$ and $b = (\sum \lambda_i)^2 / \sum \lambda_i^2$. Simulation in Ljung (1986) suggested that if ϕ is not too close to a value of one this modification gives little improvements. However, the empirical size does improve greatly if ϕ is very close to one. Battaglia (1990) considered the approximate power of Q_m. An approximate expression was also derived relating the power of Q_m and values of m.

2.4 Extension to multiplicative seasonal ARMA models

The multiplicative model is widely used in the modeling of seasonal time series. A popular example is the so-called airline model. It is so-called because Box and Jenkins (1976) first fitted the model to an airline passenger data set. It takes the form $(1-B)(1-B^{12})X_t = (1-\theta B)(1-\Theta B^{12})a_t$. In other words $W_t = (1-B)(1-B^{12})X_t$ is stationary and satisfies the moving average model $W_t = (1-\theta B)(1-\Theta B^{12})a_t = (1-\theta B - \Theta B^{12} + \theta\Theta B^{13})a_t$. Estimation of the multiplicative models is in principle the same as that for ARMA models.

Consider the general multiplicative seasonal models (SARMA) of order $(p, q) \times (P, Q)_s$ defined by

$$\Phi(B^s)\phi(B)X_t = \Theta(B^s)\theta(B)a_t \tag{2.14}$$

where a_t, $\phi(B)$ and $\theta(B)$ are defined as in (2.2),

$$\begin{aligned}\Phi(B^s) &= 1 - \Phi_1 B^s - \cdots - \Phi_P B^{P_s s} \, ,\\ \Theta(B^s) &= 1 - \Theta_1 B^s - \cdots - \Theta_Q B^{Q_s s} \, ,\end{aligned}$$

where s is the seasonal period. Let $\boldsymbol{\beta} = (\phi_1, \ldots, \phi_p, \theta_1, \ldots, \theta_q, \Phi_1, \ldots, \Phi_P, \Theta_1, \ldots, \Theta_Q)^{\mathrm{T}}$. Suppose that a_t is Gaussian with $\sigma^2 = 1$ and let $\hat{\boldsymbol{\beta}}$ be an asymptotically efficient estimator of $\boldsymbol{\beta}$. Then the asymptotic Fisher information matrix $\mathbf{I}$ is given by

$$\mathbf{I} = \left(\begin{array}{c|c} \mathbf{I}_1 & \mathbf{I}_2 \\ \hline \mathbf{I}_2^{\mathrm{T}} & \mathbf{I}_3 \end{array} \right)_{(p+q+P+Q)\times(p+q+P+Q)} ,$$

where $\mathbf{I}_1$ is given by (2.5),

$$\mathbf{I}_2 = \left(\begin{array}{c|c} \gamma_{vV}(i-js) & \gamma_{vU}(i-js) \\ \hline \gamma_{uV}(i-js) & \gamma_{uU}(i-js) \end{array} \right)^{p}_{q} \quad \begin{array}{cc} P & Q \end{array}$$

and

$$\mathbf{I}_3 = \left(\begin{array}{c|c} \gamma_{VV}((i-j)s) & \gamma_{VU}((i-j)s) \\ \hline \gamma_{UV}((i-j)s) & \gamma_{UU}((i-j)s) \end{array} \right)^{P}_{Q} \quad \begin{array}{cc} P & Q \end{array}$$

where u_t and v_t are defined in (2.6), and V_t and U_t are defined below by

$$\Phi(B^s)V_t = -a_t \ ,$$

and

$$\Theta(B^s)U_t = a_t \ .$$

Here, $\gamma_{WZ}(k) = E(W_t Z_{t+k})$, where W_t, Z_t can be one of u_t, v_t, U_t, or V_t. McLeod (1978) shows that the large sample distribution of $\hat{\boldsymbol{r}}$ is normal with mean $\mathbf{0}$ and covariance matrix

$$\mathrm{var}(\hat{\boldsymbol{r}}) = (\mathbf{1}_m - \mathbf{X}\mathbf{I}^{-1}\mathbf{X}^{\mathrm{T}})/n \qquad (2.15)$$

where

$$\mathbf{X} = \left(-\phi'_{i-j} \mid \theta'_{i-j} \mid -\Phi'_{i-js} \mid -\Theta'_{i-js} \right) m \quad \begin{array}{cccc} p & q & P & Q \end{array}$$

where Φ'_i and Θ'_i are defined by the power series expansion

$$\Phi(B^s)^{-1} = \sum_{i=0}^{\infty} \Phi'_i B^{is} \qquad (2.16)$$

and

$$\Theta(B^s)^{-1} = \sum_{i=0}^{\infty} \Theta'_i B^{is} \ .$$

As a consequence of (2.15) if $S \gg p$ and $S \gg q$, the statistics Q_m, $\tilde{Q}_m$, and Q^*_m will all have an asymptotic chi-square distribution with $m-p-q-P-Q$ degrees of freedom if the model is correct and $m \gg 0$. Note that if $\hat{\boldsymbol{\beta}}$ is obtained using an criterion other than minimizing (2.3b) it may also be possible to derive a portmanteau statistic. Examples of this are considered in later chapters.

2.5 Relation with the Lagrange multiplier test

2.5.1 The Lagrange multiplier (Score) test

Consider a statistical model involving the vector parameter $\boldsymbol{\theta} = (\boldsymbol{\theta}_1^{\mathrm{T}}, \boldsymbol{\theta}_2^{\mathrm{T}})^{\mathrm{T}}$. Suppose that we are interested in testing the null hypothesis $H_0 : \boldsymbol{\theta}_2 = \mathbf{0}$ against the alternative hypothesis $H_1 : \boldsymbol{\theta}_2 \neq 0$. Suppose further that the log-likelihood of the model $l(\boldsymbol{\theta})$ exists and can be differentiated twice continuously. Then the Lagrange or Lagrangian multiplier test for testing the above hypothesis is

$$\mathrm{LM} = \left(\frac{\partial l(\boldsymbol{\theta})}{\partial \boldsymbol{\theta}}\right)^{\mathrm{T}}_{\hat{\boldsymbol{\theta}}_1} \mathrm{E}\left(\frac{-\partial^2 l(\boldsymbol{\theta})}{\partial \boldsymbol{\theta}\partial \boldsymbol{\theta}^{\mathrm{T}}}\right)^{-1}\Bigg|_{\hat{\boldsymbol{\theta}}_1} \left(\frac{\partial l(\boldsymbol{\theta})}{\partial \boldsymbol{\theta}}\right)_{\hat{\boldsymbol{\theta}}_1} \tag{2.17}$$

where $\hat{\boldsymbol{\theta}}_1$ is the maximum likelihood estimator (MLE) of $\boldsymbol{\theta}_1$ under H_0. That is, the MLE of $\boldsymbol{\theta}_1$ assuming $\boldsymbol{\theta}_2 = 0$. Under regularity conditions $\mathrm{LM} \sim \chi_r^2$ asymptotically if H_0 is true, where r is the dimension of $\boldsymbol{\theta}_2$. The advantage of the Lagrange multiplier test over the more common likelihood ratio test is that it is not necessary to estimate the full model, i.e., the full vector of parameters $\boldsymbol{\theta}$. It is also an invariant test under the usual regularity conditions for the asymptotic normality of the MLE and is equivalent asymptotically to the likelihood ratio test. See Silvey (1959).

Consider the classical regression setup of a variable Y on two fixed regressors X_1 and X_2,

$$Y = \theta_1 X_1 + \theta_2 X_2 + a \tag{2.18}$$

where a is assumed to be i.i.d. $N(0, 1)$. A Lagrange multiplier test for $H_0 : \theta_2 = 0$ against the alternative $H_1 : \theta_2 \neq 0$ can be formed as in (2.17). Note that given observations (y_i, x_{1i}, x_{2i}), $i = 1, \ldots, n$, the log-likelihood is

$$l(\boldsymbol{\theta}) = -\frac{n}{2}\ln 2\pi - \sum_{i=1}^{n} \frac{(y_i - \theta_1 x_{1i} - \theta_2 x_{2i})^2}{2},$$

since $\partial a/\partial \theta_i = x_i$, $i = 1, 2$,

$$\frac{\partial l(\boldsymbol{\theta})}{\partial \theta_1} = \sum_{i=1}^{n}(y_i - \theta_1 x_{1i} - \theta_2 x_{2i})x_{1i} = \sum_{i=1}^{n} a_i x_{1i},$$

$$\frac{\partial l(\boldsymbol{\theta})}{\partial \theta_2} = \sum_{i=1}^{n}(y_i - \theta_1 x_{1i} - \theta_2 x_{2i})x_{2i} = \sum_{i=1}^{n} a_i x_{2i}$$

and

$$\frac{\partial^2 l(\boldsymbol{\theta})}{\partial \boldsymbol{\theta}\partial \boldsymbol{\theta}^{\mathrm{T}}} = -\begin{pmatrix} \sum x_{1i}^2 & \sum x_{1i}x_{2i} \\ \sum x_{1i}x_{2i} & \sum x_{2i}^2 \end{pmatrix}. \tag{2.19}$$

Note that under $H_0 : \partial(\boldsymbol{\theta})/\partial\theta_1|\hat{\theta}_1 = 0$ and

$$\mathrm{E}\left(\frac{-\partial^2 l(\boldsymbol{\theta})}{\partial\boldsymbol{\theta}\partial\boldsymbol{\theta}^{\mathrm{T}}}\right) = -\frac{\partial^2 l(\boldsymbol{\theta})}{\partial\boldsymbol{\theta}\partial\boldsymbol{\theta}^{\mathrm{T}}} = \sum_{i=1}^{n}(x_{1i}, x_{2i})\begin{pmatrix} x_{1i} \\ x_{2i} \end{pmatrix} .$$

Note also that

$$\left(\sum a_i x_{1i}, \sum a_i x_{2i}\right)^{\mathrm{T}} = \sum_{i=1}^{n} a_i\,(x_{1i}, x_{2i})^{\mathrm{T}} .$$

Under H_0, a_i is replaced by $\hat{a}_i$ the residual of the regression of Y on X_1. Let $\boldsymbol{X}_i = (x_{1i}, x_{2i})$, $\boldsymbol{X}^{\mathrm{T}} = (X_1^{\mathrm{T}}, \ldots, X_n^{\mathrm{T}}), \hat{\boldsymbol{a}} = (\hat{a}_1, \ldots, \hat{a}_n)^{\mathrm{T}}$. Using these results the LM test (2.17) can be written as

$$\begin{aligned} \mathrm{LM} &= \sum \hat{a}_i \boldsymbol{X}_i \left(\sum_{i=1}^{n} \boldsymbol{X}_i^{\mathrm{T}} \cdot \boldsymbol{X}_i\right)^{-1} \sum \hat{a}_i \boldsymbol{X}_i^{\mathrm{T}} \\ &= \hat{\boldsymbol{a}}^{\mathrm{T}} \boldsymbol{X}\left(\boldsymbol{X}^{\mathrm{T}}\boldsymbol{X}\right)^{-1}\boldsymbol{X}^{\mathrm{T}}\hat{\boldsymbol{a}} . \end{aligned}$$

Consider the regression

$$\hat{a}_i = \beta_1 X_{1i} + \beta_2 X_{2i} + V_i$$

which can be written

$$\begin{aligned} \hat{\boldsymbol{a}}_i &= \boldsymbol{X}_i \begin{pmatrix} \beta_1 \\ \beta_2 \end{pmatrix} + V_i \\ &= \boldsymbol{X_i \beta} + V_i \end{aligned}$$

where $\boldsymbol{\beta}^{\mathrm{T}} = (\beta_1, \beta_2)$ and V_i is i.i.d.. The coefficient of determination of the above regression is given by

$$\begin{aligned} R^2 &= \frac{\text{regression sum of squares}}{\sum_{i=1}^{n} \hat{a}_i^2} \\ &= \frac{\sum_{i=1}^{n} (\boldsymbol{X}_i\hat{\boldsymbol{\beta}})^2}{\sum_{i=1}^{n} \hat{a}_i^2} = \frac{\hat{\boldsymbol{\beta}}^{\mathrm{T}}\boldsymbol{X}^{\mathrm{T}}\boldsymbol{X}\hat{\boldsymbol{\beta}}}{\hat{\boldsymbol{a}}^{\mathrm{T}}\hat{\boldsymbol{a}}} . \end{aligned}$$

But $\hat{\beta} = (\boldsymbol{X}^{\mathrm{T}}\boldsymbol{X})^{-1}\boldsymbol{X}^{\mathrm{T}}\hat{\boldsymbol{a}}$ and hence we have

$$R^2 = \frac{\hat{\boldsymbol{a}}^{\mathrm{T}}\boldsymbol{X}(\boldsymbol{X}^{\mathrm{T}}\boldsymbol{X})^{-1}\boldsymbol{X}^{\mathrm{T}}\hat{\boldsymbol{a}}}{\hat{\boldsymbol{a}}^{\mathrm{T}}\hat{\boldsymbol{a}}} .$$

Since $\hat{\boldsymbol{a}}^{\mathrm{T}}\hat{\boldsymbol{a}}/n$ converges to 1 in probability we note that if n is sufficiently large,

$$n \cdot R^2 \cong \mathrm{LM} . \qquad (2.20)$$

Hence the Lagrange multiplier test can be computed asymptotically from n times the coefficient of determination of the regression of $\hat{\boldsymbol{a}}$ on the regressors $\partial \boldsymbol{a}/\partial \theta_1$, and $\partial \boldsymbol{a}/\partial \theta_2$. It will be seen that this result holds more generally than the setup in (2.18).

2.5.2 The LM test for ARMA time series models

Hannan (1970) has shown that it is impossible to test the null hypothesis that $\{X_t\}$ satisfies an ARMA(p, q) model against the alternative that the time series satisfies the ARMA$(p + r, q + s)$ model. However, it is possible to test the null hypothesis of ARMA(p, q) vs. either an ARMA$(p+r, q)$ or an ARMA$(p, q+s)$ alternative. In fact, the two tests will be equivalent. Let $\boldsymbol{\eta}$ be the vector of the ARMA parameters in one of these alternative models. Godfrey (1979) shows that a Lagrange multiplier test for the above can be obtained by regressing the vector of residuals $\boldsymbol{a} = (\hat{a}_1, \ldots, \hat{a}_n)^{\mathrm{T}}$ obtained under the null model on the matrix of partial derivatives $\partial \boldsymbol{a}/\partial \boldsymbol{\eta}$. Then, as in (2.20), n times the coefficient of determination of this regression is asymptotically equivalent to the Lagrange multiplier test. Note that unlike m in the Q statistics, r does not need to be large for the asymptotic chi-square distribution to be valid. Monte Carlo results in Kwan (1993) indicate that the χ^2 approximation to the distribution of the LM test may fail when the value of r is moderately large.

Newbold (1980) shows that the LM test for ARMA(p, q) vs. ARMA$(p+m, q)$ and the test based on the first m residual autocorrelations are in fact equivalent. The test based on the first m residual autocorrelations is defined by

$$S = n\,\hat{\boldsymbol{r}}^{\mathrm{T}}\,\hat{\boldsymbol{C}}^{-1}\,\hat{\boldsymbol{r}} \qquad (2.21)$$

where $\hat{\boldsymbol{C}}$ is the large sample covariance matrix of $\hat{\boldsymbol{r}}$ in (2.8) evaluated at $\hat{\boldsymbol{\beta}}$. This covariance matrix is nonsingular if m is not too large. Newbold (1980), and Ansley and Newbold (1979) advocated the use of S in model diagnostic checking based on consideration of power. Simulation results in Ljung (1986) indicate that the Newbold test, S, suffers from a size-distortion problem. Pöskitt and Tremayne (1980) considered the test of ARMA(p, q) null against the ARMA$(p + s, q + r)$ alternative based on the results of Silvey(1959, §6).

2.5.3 The Lagrange multiplier test and other goodness-of-fit tests

Goodness-of-fit tests for time series had been proposed well before the Box-Jenkins era by Quenouille (1947, 1949), Walker (1950, 1952), and

Bartlett and Diananda (1950). These tests were proposed mainly for the autoregressive models. These tests can be unified under the framework of the Lagrange multiplier test (Hosking, 1980a, Godfrey, 1979). Suppose the null hypothesis is the ARMA(p, q) model (2.1)

$$\phi(B)X_t = \theta(B)a_t \ .$$

Godfrey (1978) considered a Lagrange multiplier test for the alternative model

$$\phi(B)X_t + \sum_{i=1}^{m} \lambda_i X_{t-p-i} = \theta(B)a_t \ . \tag{2.22}$$

Hosking (1980a) considered tests for the more general alternative model

$$\phi(B)\left\{ X_t + \sum_{i=1}^{m} \lambda_i \alpha(B) X_{t-i} \right\} = \theta(B)a_t \ , \tag{2.23}$$

where $\alpha(B) = \sum_{j=0}^{\infty} \alpha_j B^j$, where all the roots of $\alpha(B)$ lie outside unit circle, and the α_is are not dependent on the λ_is. We recover the portmanteau test if $\alpha(B) \equiv 1$. Let $\boldsymbol{\lambda} = (\lambda_1, \ldots, \lambda_m)^{\mathrm{T}}$ and l be the log-likelihood of the alternative model (2.20). Let $\boldsymbol{d} = (d_1, \ldots, d_m)^{\mathrm{T}}$ where $d_i = \partial l / \partial \lambda_i$. Hosking (1980a) obtained the following general result.

Theorem 2.2 *For the hypothesis testing of models (2.1) vs. (2.23),*

$$\hat{d}_i = -n \sum_{j=0}^{\infty} \hat{\alpha}_j \ \hat{r}_{i+j} \ .$$

Under the null model (2.1) $\boldsymbol{d}$ *is asymptotically normally distributed with mean* 0 *and covariance matrix* $\boldsymbol{A} - \boldsymbol{D}\mathbf{I}^{-1}\boldsymbol{D}^{\mathrm{T}}$, *where* $\boldsymbol{A}$ *is an* $m \times m$ *matrix with* (i, j)*th*

$$a_{ij} = \langle \alpha(z)\alpha(z)^{-1} \rangle_{i-j} \ ,$$

$\mathbf{I}$ *is the information matrix of the null model (see (2.5)) and* $\boldsymbol{D} = (\boldsymbol{D}_1, \boldsymbol{D}_2)$ *where* $\boldsymbol{D}_1$ *and* $\boldsymbol{D}_2$ *are respectively* $m \times p$ *and* $m \times q$ *matrices with* (i, j)*th element*

$$\langle \alpha(z^{-1})/\phi(z) \rangle_{i-j} \quad \textit{and} \quad \langle -\alpha(z^{-1})/\theta(z) \rangle_{i-j} \ ,$$

where

$$\langle \alpha(z)\alpha(z^{-1}) \rangle_j = \sum_{k=0}^{\infty} \alpha_k \alpha_{k+j} \ , \quad j > 0 \ .$$

By letting $\alpha(z) = z^{p+q}\phi(z^{-1})\theta(z^{-1})\theta(z)/\phi(z)$, Hosking (1980a) obtained Walker's (1950) extension of Quenouille's test. When the time series is a pure autoregressive model, Quenouille's test is just $n^{-1}\hat{\boldsymbol{d}}^{\mathrm{T}}\hat{\boldsymbol{d}}$ and is asymptotically chi-squared distributed with m degrees of freedom under the null. Hosking (1980a) also obtained other types of goodness-of-fit tests by allowing different $\alpha(z)$ that are functions of $\phi(z)$ and $\theta(z)$. As was considered in Hosking (1978) this unification of goodness-of-fit tests may also be done based on the result of Durbin (1970). The Hosking's result is useful when one has specific alternatives in mind and in general this should provide somewhat more powerful tests than the portmanteau test, which has better power with the alternative model (2.22) but may not be as powerful with other alternatives. This result also suggests that the portmanteau tests are not just pure significance tests but can be viewed as Lagrangian multiplier tests under appropriate alternatives. Godfrey and Tremayne (1988) gave a review on various tests for univariate time series. Godolphin (1978, 1980) considered some alternative testing procedures for univariate ARMA models.

2.6 A test based on the residual partial autocorrelation test

Monti (1994) proposed a portmanteau test similar to (2.12) using the residual partial autocorrelations $\hat{\pi}_k$, $k = 1, \ldots, m$. It was shown that $\hat{\pi} = (\hat{\pi}_1, \ldots, \hat{\pi}_m)^{\mathrm{T}}$ and $\hat{\boldsymbol{r}}$ are asymptotically equivalent, *viz.*,

$$\hat{\boldsymbol{\pi}} = \hat{\boldsymbol{r}} + O_p(n^{-1}) \ .$$

Note that $\hat{\boldsymbol{\pi}}$ can be obtained from $\hat{\boldsymbol{r}}$ using the Durbin-Levinson algorithm (Box and Jenkins, 1976, p.82). Hence, the statistic

$$\tilde{Q}_m(\hat{\pi}) = n(n+2) \sum_{k=1}^{m} \hat{\pi}_k^2/(n-k) \qquad (2.24)$$

is asymptotically distributed as χ^2_{m-p-q} if the fitted ARMA model is adequate. Simulation experiments reported in Monti (1994) suggested that the performance of $\tilde{Q}_m(\hat{\pi})$ is comparable to that of $\tilde{Q}_m$ and better if the order of the moving average is understated. On the other hand, $\tilde{Q}_m$ is more powerful if the order of the autoregressive part is understated. Monte Carlo results in Kwan and Wu (1997) suggested that the performance of the two is very similar.

2.7 A test based on the residual correlation matrix test

Let the residual correlation matrix $\hat{\boldsymbol{R}}_m$, of order m, be given by

$$\hat{\boldsymbol{R}}_m = \begin{bmatrix} 1 & \hat{r}_1 & \cdots & \cdots & \hat{r}_m \\ \hat{r}_1 & 1 & \cdots & \cdots & \hat{r}_{m-1} \\ & & \ddots & & \\ & & & \ddots & \\ \hat{r}_m & \hat{r}_{m-1} & \cdots & \cdots & 1 \end{bmatrix} . \tag{2.25}$$

Peña and Rodriǵuez (2002) proposed a portmanteau test $\hat{D}_m$ based on the determinant $|\hat{\boldsymbol{R}}_m|$ of $\hat{R}_m$. It is defined as

$$\hat{D}_m = n\big[1 - |(\hat{\boldsymbol{R}}_m)|^{1/m}\big] . \tag{2.26}$$

Now it may be shown that

$$|\hat{\boldsymbol{R}}_m| = |\hat{\boldsymbol{R}}_{m-1}|(1 - \hat{R}_m^2) \tag{2.27}$$

where $\hat{R}_m^2 = \hat{\boldsymbol{r}}^{\mathrm{T}} \hat{\boldsymbol{R}}_{m-1}^{-1} \hat{\boldsymbol{r}}$ is just the square of the multiple correlation coefficient in the regression of $\hat{a}_t$ on $\hat{a}_{t-1}, \ldots, \hat{a}_{t-m}$. Iterating (2.27) gives

$$|\hat{\boldsymbol{R}}_m| = (1 - \hat{R}_1^2) \cdots (1 - \hat{R}_m^2) . \tag{2.28}$$

Hence $|\hat{\boldsymbol{R}}_m|^{1/m}$ can be interpreted as the geometric mean of the product in (2.28). Alternatively it is also well known that (Hannan, 1970, p.22)

$$|\hat{\boldsymbol{R}}_m|^{1/m} = \prod_{i=1}^{m} (1 - \hat{\pi}_i^2)^{(m+1-i)/m}$$

where $\hat{\pi}_i$ is the i-th residual partial autocorrelation. Hence $|\hat{R}_m|^{1/m}$ can be seen as a weighted function of $\hat{\pi}_i^2$, $i = 1, \ldots, m$. Peña and Rodriǵuez (2002) showed that if the model is adequate $\hat{D}_m$ is asymptotically distributed as $\sum_{i=1}^{m} \lambda_i \chi_{1,i}^2$, where $\chi_{1,i}^2$ are independent chi-square random variables with one degree of freedom and λ_i are the eigenvalues of $\big(\mathbf{1}_m - \mathbf{X}\,\mathbf{I}^{-1}\,\mathbf{X}^{\mathrm{T}}\big)\boldsymbol{W}_m$, where $\boldsymbol{W}_m$ is a diagonal matrix with i-th diagonal element $W_i = (m - i + 1)/m$, $i = 1, \ldots, m$. In practice, it has been suggested that the distribution of $\hat{D}_m$ be approximated by a gamma distribution $G(\alpha, \beta)$ with parameters $\alpha = b/2$, $\beta = a/2$, where $a = \sum \lambda_i^2 / \sum \lambda_i$ and $b = (\sum \lambda_i)^2 / \sum \lambda_i^2$. Peña and Rodriǵuez actually considered a modification D_m of $\hat{D}_m$ by replacing $\hat{r}_k$ in $\hat{\boldsymbol{R}}_m$ with $\tilde{r}_k$, where $\tilde{r}_k = [(n + 2)/(n - k)]^{1/2} \hat{r}_k$. Simulations in Peña and Rodriǵuez (2002) suggested that D_m has better power than either $\tilde{Q}_m$ or $\tilde{Q}_m(\hat{\pi})$. They also applied D_m to the squared residuals for checking the assumption of linearity. A recent study by Kwan and Wu (2003), however, suggested that there could be serious size distortation with the Peña–Rodriǵuez test.

2.8 Extension to periodic autoregressions

A useful class of models for hydrological time series has been the periodic autoregressive (PAR) model. Suppose that there are s seasonal periods in a year and there are n years of data available. The time index t may be parameterized as $t = (r-1)s + v = t(r,v)$, where $r = 1,\ldots,n$ and $v = 1,\ldots,s$. Denote by μ_v the mean of X_t for the v-th seasonal period. The lag l autocorrelation for the v-th season is defined by

$$\gamma_{l,v} = \text{cov}(X_{t(r,v)}, X_{t(r,v)-l}) \,, \tag{2.29}$$

where cov(,) is the covariance operator. A PAR model of order $(p_1,\ldots,$ $p_s)$ is defined by

$$X_{t(r,v)} = \mu_v + \sum_{i=1}^{pv} \phi_{i,v}(X_{t(r,v)-i} - \mu_{v-i}) + a_{t(r,v)} \tag{2.30}$$

where $a_{t(r,v)}$ is white noise with mean 0 and variance σ_v^2. Note that the distribution of $a_{t(r,v)}$ is different for different seasons. The periodic time series $X_{t(r,v)}$ has a moving average representation (Troutman, 1979),

$$X_{t(r,v)} = \mu_v + \sum_{i=0}^{\infty} \psi_{i,v} a_{t(r,v)-i} \tag{2.31}$$

where $\psi_{0,v} = 1$, $\psi_{i,v} = 0$ if $i \le 0$ and

$$\psi_{i,v} = \sum_{j=0}^{p_m} \phi_{j,v}\psi_{i-j,v-j}, \quad i \ge 1 \,.$$

Estimation of the PAR was discussed in Pagano (1978) and Newton (1982). Extension to periodic ARMA (PARMA) models was considered by Vecchia (1985). Exact likelihood estimation for the PARMA model was considered in Li and Hui (1988). Define the residuals from the PAR fitted to $X_{t(r,v)}$ by $\hat{a}_{t(r,v)}$. The lag l residual autocorrelation for the v-th season is given by

$$\hat{r}_{l,v} = \frac{\sum_r \hat{a}_{t(r,v)}\hat{a}_{t(r,v)-l}}{[\sum_r \hat{a}^2_{t(r,v)} \sum_r \hat{a}^2_{t(r,v)-l}]^{1/2}} \,. \tag{2.32}$$

Let $\hat{\boldsymbol{r}}_v = (\tilde{r}_{1,v},\ldots,\hat{r}_{m,v})^{\mathrm{T}}$. Then it can be shown that $\sqrt{n}\hat{\boldsymbol{r}}_v$ is asymptotically normal with mean zero and

$$\text{var}(\hat{\boldsymbol{r}}_v) = \mathbf{1}_m - \mathbf{X}_v \mathbf{I}_v^{-1} \mathbf{X}_v^{\mathrm{T}}$$

where $\mathbf{X}_v$ has (i,j)th entry $-\psi_{i-j,v}\sigma_{v-j}/\sigma_v$, $1 \le i \le m$, $1 \le j \le p_v$ and $\mathbf{I}_v$ is the information matrix of the autoregressive parameters for the v-th season. It can also be shown that $\sqrt{n}\hat{\boldsymbol{r}}_v$ and $\sqrt{n}\hat{\boldsymbol{r}}_{v'}$, $v \ne v'$

are asymptotically independent which implies that a portmanteau test can be carried out individually for each season. Based on these results McLeod (1994) suggested the following modified portmanteau statistic

$$\tilde{Q}_{L,v} = \sum_{l=1}^{m} \frac{\hat{r}_{l,v}^2}{\{\mathrm{var}(r_{l,v})\}} \tag{2.33}$$

where $\mathrm{var}(r_{l,v}) = \{n - [(l - v + s)/s]\}/n^2$ and $[\cdot]$ denotes the integer part function. If the model is adequate $\tilde{Q}_{L,v}$ will be asymptotically distributed as a chi-square variable with $m - p_v$ degrees of freedom. Using simulation, McLeod (1994) demonstrated that $\tilde{Q}_{L,v}$ has good size properties with n as low as 50. The treatise by Hipel and McLeod (1994, Ch.14) contains a full exposition on the modeling of PARMA models.

CHAPTER 3

The multivariate linear case

3.1 The vector ARMA model

In many applications we would like to model the relationship between say, time series $X_{1t}, X_{2t}, \ldots, X_{lt}$, where l is an integer greater than one. By writing $\boldsymbol{X}_t = (X_{1t}, X_{2t}, \ldots, X_{lt})^{\mathrm{T}}$ model (2.1) can be easily extended to handle the multivariate situation. Let $\boldsymbol{a}_t = (a_{1t}, a_{2t}, \ldots, a_{lt})^{\mathrm{T}}$ and $\boldsymbol{\Phi}_i$, $i = 1, \ldots, p$; $\boldsymbol{\Theta}_j$, $j = 1, \ldots, q$, be $l \times l$ coefficient matrices. Then the vector (multivariate) autoregressive moving average (VARMA(p, q)) model is defined by

$$\boldsymbol{X}_t - \boldsymbol{\Phi}_1 \boldsymbol{X}_{t-1} - \cdots - \boldsymbol{\Phi}_p \boldsymbol{X}_{t-p} = \boldsymbol{a}_t - \boldsymbol{\Theta}_1 \boldsymbol{a}_{t-1} - \cdots - \boldsymbol{\Theta}_q \boldsymbol{a}_{t-q} \quad (3.1)$$

where $\boldsymbol{a}_t$ is assumed to be an l dimensional white noise process. That is, $\boldsymbol{a}_t$ is uncorrelated over time with mean zero and covariance matrix $\boldsymbol{\Delta}$. A constant vector $\boldsymbol{\Theta}_0$ may also be added to the r.h.s. of (3.1). In terms of the backshift operator B (3.1) can be written

$$\boldsymbol{\Phi}(B) \boldsymbol{X}_t = \boldsymbol{\Theta}(B) \boldsymbol{a}_t \quad (3.2)$$

where $\boldsymbol{\Phi}(B) = \mathbf{1}_l - \boldsymbol{\Phi}_1 B - \cdots - \boldsymbol{\Phi}_p B^p$ and $\boldsymbol{\Theta}(B) = \mathbf{1}_l - \boldsymbol{\Theta}_1 B - \cdots - \boldsymbol{\Theta}_q B^q$, where $\mathbf{1}_l$ is the $l \times l$ identity matrix. For the process $\boldsymbol{X}_t$ to be stationary it is required that all roots of $\det\{\boldsymbol{\Phi}(B)\}$ have modulus greater than one or equivalently lie outside the unit circle; $\det\{\cdot\}$ here denotes the determinant function. Similarly for invertibility it is required that all roots of $\det\{\boldsymbol{\Theta}(B)\}$ lie outside the unit circle. For identifiability it is required that $\boldsymbol{\Phi}(z)$ and $\boldsymbol{\Theta}(z)$ have no common left factors and that the matrix $[\boldsymbol{\Phi}_p : \boldsymbol{\Theta}_q]$ is of full rank (Hannan, 1969; Granger and Newbold, 1986). When $q = 0$, we have the pure vector autoregressive process (VAR)

$$\boldsymbol{X}_t - \boldsymbol{\Phi}_1 \boldsymbol{X}_{t-1} - \cdots - \boldsymbol{\Phi}_p \boldsymbol{X}_{t-p} = \boldsymbol{a}_t \quad (3.3)$$

and when $p = 0$, we have a pure vector moving average process (VMA),

$$\boldsymbol{X}_t = \boldsymbol{a}_t - \boldsymbol{\Theta}_1 \boldsymbol{a}_{t-1} - \cdots - \boldsymbol{\Theta}_q \boldsymbol{a}_{t-q} \,. \quad (3.4)$$

The Box-Jenkins methodology for fitting univariate ARMA models can be extended naturally to stationary VARMA models. In a pure VMA(q)

initial identification of the model order q can be made using the sample autocorrelation matrix R_k of $\boldsymbol{X}_t$, which is defined analogously as in the univariate case. Let the length of realization of $\boldsymbol{X}_t$ be n. The lag k sample autocovariance matrix $\boldsymbol{C}_k$ for a realization of length n is given by

$$\boldsymbol{C}_k = \frac{1}{n} \sum_{t=k+1}^{n} (\boldsymbol{X}_t - \bar{\boldsymbol{X}})(\boldsymbol{X}_{t-k} - \bar{\boldsymbol{X}})^{\mathrm{T}} \tag{3.5}$$

where $\bar{\boldsymbol{X}} = n^{-1} \cdot \sum_{t=1}^{n} \boldsymbol{X}_t$. Let $\boldsymbol{D}$ be the diagonal matrix with the i-th diagonal element the square root of $n^{-1} \sum_{i=1}^{n} (X_{it} - \bar{X}_{it})^2$. Then the lag k sample autocorrelation matrix $\boldsymbol{R}_k$ is given by

$$\boldsymbol{R}_k = \boldsymbol{D}^{-1} \boldsymbol{C}_k \boldsymbol{D}^{-1} , \quad k \geq 1 . \tag{3.6}$$

Let $\boldsymbol{\rho}_k = E(\boldsymbol{D}^2)^{-1/2} E(\boldsymbol{C}_k) E(\boldsymbol{D}^2)^{-1/2}$ then $\boldsymbol{\rho}_k \equiv 0$ for $k > q$ when $\boldsymbol{X}_t$ follows a VMA(q) process. This implies that $\boldsymbol{R}_k \cong 0$ for n large and $k > q$. As in the univariate situation this property enables us to identify q empirically. For the VAR(p) process a vector partial correlation coefficient at lag k may be defined using the working autoregression

$$\boldsymbol{X}_t = \boldsymbol{\Phi}_{11} X_{t-1} + \cdots + \boldsymbol{\Phi}_{kk} \boldsymbol{X}_{t-k} + \varepsilon_t , \quad k \geq 1 \tag{3.7}$$

where ε_t is just an l dimensional residual. Note that if $E(\boldsymbol{X}_t) \neq 0$ then without loss of generality we refer to the centered time series also as $\boldsymbol{X}_t$. The coefficient $\boldsymbol{\Phi}_{kk}$ of $\boldsymbol{X}_{t-k}$ can be taken to be the vector partial autocorrelation of $\boldsymbol{X}_t$ at lag k. Like its univariate counterpart $\boldsymbol{\Phi}_{kk} \equiv 0$ if $k > p$ and hence its empirical counterpart based on n observations $\hat{\boldsymbol{\Phi}}_{kk} \cong 0$ for n large and $k > p$. This property can be used to identify the autoregressive order p. More elaborated model building strategies can be found in the relevant chapter by Tiao in Peña, Tiao, and Tsay (2001). Estimation of parameters is then facilitated by assuming $\boldsymbol{a}_t$ to be Gaussian so that an (approximate) maximum likelihood estimator (MLE) procedure can be used. The initial estimates of p and q can be refined at the model diagnostic checking stage based on the residual autocorrelation matrices $\hat{\boldsymbol{R}}_k$ of the residuals $\hat{\boldsymbol{a}}_t$. An overall portmanteau test for testing whether the residuals $\hat{\boldsymbol{a}}_t$ are approximately white noise has been derived (in the VAR(p) case) by Chitturi (1974), and in the general VARMA(p, q) case by Hosking (1980b) and Li and McLeod (1981). Basically it was shown that in the general VARMA(p, q) case the statistic

$$Q(m) = n \cdot \sum_{k=1}^{m} \mathrm{tr}(\hat{\boldsymbol{C}}_k^{\mathrm{T}} \hat{\boldsymbol{C}}_0^{-1} \hat{\boldsymbol{C}}_k \hat{\boldsymbol{C}}_0^{-1}) , \tag{3.8}$$

is asymptotically chi-squared distributed with degrees of freedom $l^2(m - p - q)$ if the model is adequate and $n \gg m \gg 0$. $\hat{\boldsymbol{C}}_k$ is the lag k residual autocovariance matrix of $\hat{\boldsymbol{a}}_t$ and $\mathrm{tr}(\cdot)$ denotes the trace function for matrices. Unlike (3.6) above, Chitturi (1974) used a definition of $\hat{\boldsymbol{R}}_k = \hat{\boldsymbol{C}}_k \hat{\boldsymbol{C}}_0^{-1}$ while Hosking (1980b) used $\hat{\boldsymbol{R}}_k = \boldsymbol{L}^{\mathrm{T}} \hat{\boldsymbol{C}}_k \boldsymbol{L}$ where $\boldsymbol{L}\boldsymbol{L}^{\mathrm{T}} = \hat{\boldsymbol{C}}_0^{-1}$. Hosking (1981b) shows that all three forms of $\hat{R}_k$ give rise to the same $Q(m)$ statistic (3.8). As in the univariate case modification to (3.8) in the finite sample case is required. Hosking (1980b) considered the modified statistic

$$\tilde{Q}(m) = n^2 \sum_{k=1}^{m} \frac{1}{n-k} \mathrm{tr}(\hat{\boldsymbol{C}}_k^{\mathrm{T}} \hat{\boldsymbol{C}}_0^{-1} \hat{\boldsymbol{C}}_k \boldsymbol{C}_0^{-1}) \tag{3.9}$$

which is similar to the adjustment used in the univariate Ljung-Box statistic while Li and McLeod (1981) suggested using

$$Q^*(m) = n \sum_{k=1}^{m} \mathrm{tr}(\hat{\boldsymbol{C}}_k^{\mathrm{T}} \hat{\boldsymbol{C}}_0^{-1} \hat{\boldsymbol{C}}_k \boldsymbol{C}_0^{-1}) + \frac{l^2 m(m+1)}{2n} . \tag{3.10}$$

Note that the statistics $Q^*(m)$ and $Q(m)$ have the same variance. One criticism of the Ljung-Box adjustment (Davies, Triggs, and Newbold, 1977) which also applies to $\tilde{Q}(m)$ is that the variance of the statistic could be much larger than that of a chi-squared distribution with $l^2(m-p-q)$ degrees of freedom and thus resulting in a test that could be too sensitive. To check the effectiveness of (3.10), Li and McLeod (1981) considered the first order bivariate autoregressive model with $n = 200$, generated by a zero mean Gaussian $\boldsymbol{a}_t$ process with covariance matrix

$$\boldsymbol{\Delta} = \begin{pmatrix} 1 & \alpha \\ \alpha & 1 \end{pmatrix},$$

where $\alpha = \pm 0.25, \pm 0.5, \pm 0.75$, and $\boldsymbol{\phi}_1 = \boldsymbol{A}, \boldsymbol{B}, \boldsymbol{C}$ with

$$\boldsymbol{A} = \begin{pmatrix} -0.2 & 0.3 \\ -0.6 & 1.1 \end{pmatrix}, \ \boldsymbol{B} = \begin{pmatrix} 0.4 & 0.1 \\ -1.0 & 0.5 \end{pmatrix} \text{ and } \boldsymbol{C} = \begin{pmatrix} -1.5 & 1.2 \\ -0.9 & 0.5 \end{pmatrix}.$$

One thousand independent samples were simulated in each case and the portmanteau statistics defined in (3.8) and (3.10) were calculated with $m = 20$. The 5% empirical significance levels for Q_{20} and Q_{20}^*, shown in Table 3.1, are defined as the proportion of times that the statistic exceeds the upper 5% of χ^2_{76}. It can be seen that the modified portmanteau test (3.10) provides a significant improvement.

Ledolter (1983) conducted some more simulation experiments on the three portmanteau statistics. In general, both $\tilde{Q}$ and Q^* provide considerable improvements over Q in terms of size at $n = 100$ and 200 with $m = 15$ and 20, respectively.

The Lagrangian multiplier test framework mentioned in Chapter 2 can also be extended to the vector case. This was largely the work of Hosking (1981a). As in the univariate case the portmanteau test (3.8) can be

Table 3.1 *Empirical significance of the portmanteau tests at the 5% level in % (Li and McLeod 1981). ©1981 The Royal Statistical Society, reproduced with the permission of Blackwell Publishing*

	A		***B***		***C***	
α	Q_{20}	Q^*_{20}	Q_{20}	Q^*_{20}	Q_{20}	Q^*_{20}
0.25	32	58	29	57	28	61
−0.25	31	56	28	52	27	55
0.5	33	56	27	56	28	62
−0.5	30	56	26	64	32	66
0.75	33	57	22	57	26	60
−0.75	36	73	26	70	36	74

derived as a special type of alternative to the fitted VARMA(p, q) model. This general alternative H_1 is of the form

$$\boldsymbol{a}_t + \sum_{r=1}^{m} \boldsymbol{E}(B)\,\boldsymbol{\Lambda}_r\,\boldsymbol{F}(B)\,\boldsymbol{a}_{t-r} = \boldsymbol{\varepsilon}_t \tag{3.11}$$

where $\boldsymbol{E}(B)$ and $\boldsymbol{F}(B)$ are functions of $\boldsymbol{\Phi}_i$ and $\boldsymbol{\Theta}_j$, and $\boldsymbol{\Lambda}_r$, $r = 1, \ldots, m$ are additional parameters independent of $\boldsymbol{E}(B)$ and $\boldsymbol{F}(B)$ and $\boldsymbol{\varepsilon}_t$ is white noise. The roots of $\det\{\boldsymbol{E}(B)\}$ and $\det\{\boldsymbol{F}(B)\}$ are assumed to lie outside the unit circle for the time series to be stationary. Hosking (1981a) gives some additional conditions on $\boldsymbol{E}(B)$ and $\boldsymbol{F}(B)$. When a pure VAR(p) is considered the statistic (3.8) corresponds to the case $\boldsymbol{E}(B) = \boldsymbol{F}(B) = \mathbf{1}_l$. This is equivalent to testing the alternative model VARMA(p, m) against the VAR(p) null. A multivariate Quenouille test is possible (Hosking, 1981a). A new multivariate extension of the univariate Quenouille's test is also obtained by Hosking. See also Ledolter (1983). A stepwise testing procedure using the Lagrangian multiplier test has been developed by Pötscher (1983). Pöskitt and Tremayne (1982) considered Lagrangian multiplier tests under a Pitman sequence of alternatives. The distribution of the portmanteau test for nonstationary multivariate ARMA models has been considered by T.M. Tang in a Hong Kong University of Science and Technology M.Phil. thesis, 2003. Extension of (3.8) to structural parameterization in vector autoregressive models was considered in Ahn (1988).

3.2 Granger causality tests

3.2.1 Causality

The problem of causal relationship has been a fascinating subject for both philosophers and statisticians for centuries. In statistics, when a student first comes across simple correlation analysis, he is usually cautioned that a significant cross-correlation does not necessarily imply a cause and effect type relationship. On the other hand, it is difficult to define clearly what causality means. Granger (1969, 1980a) proposed a framework to study causal relationships in time series analysis.

For simplicity, consider as in Granger (1980a), a "universe", or equivalently an information set, in which all variables are measured at pre-specified time points and equally spaced intervals. Let F_n be the set of all knowledge in that universe up to and including time n. If Y_t is a variable in that universe, denote by $F_n - Y_n$ the set of all knowledge of that universe at time n excluding past and present values of Y_t. It seems natural to follow Granger (1980a) in assuming the following two axioms:

Axiom A. The past and present may cause the future but not conversely.
Axiom B. F_n contains no redundant information in the sense that if a variable Z is functionally related to one or more other variables in a deterministic fashion, then Z would be excluded from F_n.

Suppose at $t = n$, X_{n+1} is a random variable. Then a variable Y_n is said to cause X_{n+1} if for some set A

$$\text{Prob}\,(X_{n+1} \in A | F_n) \neq \text{Prob}\,(X_{n+1} \in A | F_n - Y_n) \; .$$

That is, Y_n causes X_{n+1} provided that the probability statement about X_{n+1} is altered with the use of Y_n as an additional piece of information.

Granger's definition above is similar in spirit to that of Suppes (1979), namely, an event B_t, (occurring at time t') is a prima facie cause of the event E_t if (i) $t' < t$, (ii) $\text{Prob}(B_{t'}) > 0$, and (iii) $\text{Prob}(E_t|B_{t'}) > P(E_t)$. The readers are referred to Granger (1980a) for more detailed discussion.

It is clear that Granger's definition is not operational in actual practice. However, an operational definition of causality between two time series can be defined in terms of predictability (Granger, 1969). A variable X is said to cause another variable Y, with respect to a given universe or information set that includes X and Y, if present Y can be better predicted by using past values of X than by not doing so, all other relevant information (including the past of Y) in the universe being used in either case. In this definition of causality it is not required that the variables involved satisfy a linear system. However, if the variables actually satisfy a linear system then comparisons of linear predictions are called for. Suppose X_t and Y_t are two time series. Let A_t, for $t = 0, \pm 1, \pm 2, \ldots$, be the given information set that includes at least X_t and Y_t. Let $\bar{A}_t = \bigcup_{s<t} A_s$, $\tilde{A}_t = \bigcup_{s \leq t} A_s$ and similarly define information sets $\bar{X}_t$, $\tilde{X}_t$, $\bar{Y}_t$, and $\tilde{Y}_t$. Given the information set A_t, consider the minimum mean square error one step ahead predictor of Y_t and denote the resulting mean square error by $\sigma^2(Y|A_t)$. According to Granger (1969) X causes Y (in the universe $\bar{A}_t$) if

$$\sigma^2\left(Y|\bar{A}_t\right) < \sigma^2\left(Y|\bar{A}_t - \bar{X}_t\right)$$

while X causes Y instantaneously (in the universe $\bar{A}_t \cup \tilde{X}_t$) if

$$\sigma^2\left(Y|\bar{A}_t, \tilde{X}_t\right) < \sigma^2\left(Y|\bar{A}_t\right) \ .$$

Causality from Y to X can be defined in the same way. Feedback occurs when X causes Y and Y also causes X. Note that Wiener (1956) originally formulated such a definition of causality between two time series, whereas Pierce and Haugh (1977) expanded upon the work of Granger (1969) and gave a comprehensive survey regarding research on causality in temporal systems. For simplicity we consider the case of two time series $\{X_t\}$ and $\{Y_t\}$.

Let X_t and Y_t be such that

$$\phi_X(B)\left(X_t - \mu_X\right) = \theta_X(B)u_t \tag{3.12a}$$

and

$$\phi_Y(B)\left(Y_t - \mu_Y\right) = \theta_Y(B)v_t \tag{3.12b}$$

where μ_X is the theoretical mean of the X_t series; $\phi_X(B) = 1 - \phi_{X,1}B - \phi_{X,2}B^2 - \cdots - \phi_{X,p_X}B^{p_X}$ is the autoregressive (AR) operator of order p_X such that the roots of the characteristic function $\phi_X(B)$ lie outside the unit circle; $\theta_X(B) = 1 - \theta_{X,1}B - \theta_{X,2}B^2 - \cdots - \theta_{X,q_X}B^{q_X}$ is the moving average (MA) operator of order q_X such that the roots of $\theta_X(B)$ lie outside the unit circle for invertibility, and u_t is white noise (also called innovation or disturbance) that has a mean of zero and variance of σ_u^2; similarly for μ_Y, $\phi_Y(B)$, $\theta_Y(B)$, and v_t, respectively.

Let the cross-correlation function (CCF), at lag k between the u_t and v_t series in (3.12a) and (3.12b), be defined as

$$\rho_{uv}(k) = \frac{E(u_t v_{t+k})}{\left[E(u_t^2)E(v_t^2)\right]^{1/2}} \ . \tag{3.13}$$

Pierce and Haugh (1977) explained that there are many possible types of causal interactions between X and Y which can be characterized by the properties of $\rho_{uv}(k)$. Using the results of Pierce (1977, Table 1) the causal relationships from X to Y are categorized according to the restrictions on $\rho_{uv}(k)$ in Table 3.2. The cases for causal relationships from Y to X can be defined similarly using $\rho_{uv}(k)$ for $k \leq 0$. See also Hipel et al. (1985). Price (1979) and also Pierce and Haugh (1979) pointed out that any of the relationships in Table 3.2 which involve instantaneous causality are only valid when there is no feedback (feedback occurs when Y also causes X). The entries in Table 3.2 are self-explanatory. For example, unidirectional causality from X to Y corresponds to $\rho_{uv}(k) \neq 0$ for some $k > 0$, $\rho_{uv}(k) = 0$ for all $k < 0$, but $\rho_{uv}(0)$ may either be zero or else have some nonzero value between -1 and 1. Where X does not cause Y at all, instantaneous causality does not exist between X and Y since $\rho_{uv}(0) = 0$.

Table 3.2 *Causal relationships between two variables as characterized by* $\rho_{uv}(k)$

Relationship	Restrictions on $\rho_{uv}(k)$
X causes Y	$\rho_{uv}(k) \neq 0$ for some positive k
Instantaneous Causality	$\rho_{uv}(0) \neq 0$
X causes Y but not instantaneously	$\rho_{uv}(k) \neq 0$ for some positive k and $\rho_{uv}(0) = 0$
X does not cause Y	$\rho_{uv}(k) = 0$ for all positive k
X does not cause Y at all	$\rho_{uv}(k) = 0$ for all non-negative k
Unidirectional causality from X to Y	$\rho_{uv}(k) \neq 0$ for some $k > 0$ and $\rho_{uv}(k) = 0$ for either (a) all $k < 0$ or (b) all $k \leq 0$
X and Y are only related instantaneously	$\rho_{uv}(0) \neq 0$ and $\rho_{uv}(k) = 0$ for all $k \neq 0$
X and Y are uncorrelated	$\rho_{uv}(k) = 0$ for all k

In practice the estimated CCF $r_{\hat{u}\hat{v}}(k)$ of the model residuals is used in place of the CCF of u_t and v_t to ascertain which $\rho_{uv}(k)$s are significantly different from zero. Under the null that u_t and v_t are uncorrelated it can be shown that $\sqrt{n}\,\boldsymbol{r}_{\hat{u}\hat{v}} = \sqrt{n}(r_{\hat{u}\hat{v}}(1), \ldots, r_{\hat{u}\hat{v}}(S))^{\mathrm{T}}$ is asymptotically normally distributed with mean zero and covariance matrix $\mathbf{1}_S$. Consequently a portmanteau test of independence can be based on the statistic

$$P(S) = n^2 \sum_{k=1}^{S} r_{\hat{u}\hat{v}}^2(k) \,/(n-k) \tag{3.14}$$

which is asymptotically chi-squared with degrees of freedom S. Usually

S is of the order $n/4$. If $r^2_{\hat{u}\hat{v}}(0)$ is included in (3.14) then $P(S)$ has degrees of freedom $S+1$. See Haugh (1976).

McLeod (1979) considers the asymptotic distribution of $\boldsymbol{r}_{\hat{u}\hat{v}}$ under the assumption that the processes u_t and v_t are correlated. We will discuss this result in greater detail in Chapter 4. An important special case is where $\rho_{uv}(k) = \rho$ if $k = 0$, and zero otherwise. Such time series have only a contemporaneous correlation through their noise process and are akin to the so-called seemingly uncorrelated regression situation in econometrics. Even in this simple case (3.14) has to be modified as

$$P'(S) = n \cdot \boldsymbol{r}^{\mathrm{T}}_{\hat{u}\hat{v}} P_1^{-1} \boldsymbol{r}_{\hat{u}\hat{v}} \tag{3.15}$$

where $P_1 = \mathbf{1}_s - \rho^2 \mathbf{X}\mathbf{I}_1\mathbf{X}^{\mathrm{T}}$, $\mathbf{I}_1$ is the information matrix for model (3.12a) and $\mathbf{X}$ is evaluated using (2.7) under (3.12a). $P'(S)$ is asymptotically chi-squared with degrees of freedom S under the null of no cross-correlation between u_t and v_t. The tests $P(S)$ and $P'(S)$ for some large S can thus be viewed as tests of the null hypothesis of no Granger causality. Alternatively suppose $\boldsymbol{z}_t = (X_t, Y_t)^{\mathrm{T}}$ can be modeled by a bivariate VAR (3.3) of order S. In the case of no feedback the $P(S)$ test is also equivalent to testing the null hypothesis

$$H_0 : \phi_{1,21} = \phi_{2,21} \cdots = \phi_{S,21} = 0$$

for some large S, where $\phi_{i,21}$, $i = 1, 2, \ldots, S$, is the lower left-hand corner entry of $\boldsymbol{\Phi}_i$. See Granger and Newbold (1986) for the testing of causality using the VAR framework.

3.2.2 *Prewhitening and power*

Cross-correlation analysis of the residuals of univariate time series models for testing the independence of two time series was first suggested by Fisher (1921), in the context of orthogonal polynomial trend models. Jenkins and Watts (1968, p.339) proposed the same approach using univariate autoregressive models for the two time series. Further developments of the residual cross-correlation approach have been considered by several researchers (Haugh, 1976; Haugh and Box, 1977; Pierce, 1977; Pierce and Haugh, 1977; Sims, 1977; McLeod, 1979). The simplicity and intuitive appeal of this test for independence has been stressed by many of the authors. Nevertheless, arguments for the residual cross-correlation approach can be made even more convincing if the power function of an associated test is computed for a plausible alternative hypothesis, and compared with the power function of tests based upon other approaches.

In one of the simplest possible types of dependence between two time series, the only nonzero cross-correlations of the innovation series which generate them occur between innovations corresponding to the same lag over time. Autoregressive moving average models where dependence is of this "causality at one point only" nature appear to be suitable for the empirical description of the relationships between economic time series (Pierce, 1977).

As in McLeod and Li (1983) and Li (1981), consider time series x_t and y_t $(t = 1, \ldots, n)$ which are generated by the zero mean autoregressive models

$$X_t = \phi_X X_{t-1} + a_t$$

and

$$Y_t = \phi_Y Y_{t-1} + b_t \tag{3.16}$$

where $a_t \sim NID(0, \sigma_a^2)$, $b_t \sim NID(0, \sigma_b^2)$, $|\phi_x| < 1$ and $|\phi_y| < 1$. Suppose that the innovation series a_t and b_t are jointly normal and that the cross-correlation function between a_t and b_t is

$$\begin{aligned} \rho_{ab}(j) = \rho \ , & \qquad j = k \\ 0 \ , & \qquad j \neq k \end{aligned} \tag{3.17}$$

where $\rho_{ab}(j) = \text{cov}(a_t, b_{t+j})/(\sigma_a \sigma_b)$ and $l = 0, \pm 1, \pm 2, \ldots$.

The parameter ρ measures the degree of the dependence between the time series X_t and Y_t. Thus, for testing the independence of X_t and Y_t, the null hypothesis

$$H_0 : \rho = 0$$

can be tested against the alternative

$$H_1 : \rho \neq 0 \ .$$

In the univariate residual cross-correlation approach, the first step is to obtain univariate estimates which are asymptotically efficient for the parameters ϕ_X and ϕ_Y. Denote realized values of X_t and Y_t by x_t and y_t, respectively. Such estimators are given by

$$\hat{\phi}_X = \sum_{t=2}^{n} x_t x_{t-1} \Big/ \sum_{t=2}^{n} x_t^2 \tag{3.18}$$

and

$$\hat{\phi}_Y = \sum_{t=2}^{n} y_t y_{t-1} \Big/ \sum_{t=2}^{n} y_t^2 \ .$$

Then, the residual cross-correlation

$$r_{\hat{a}\hat{b}}(j) = \frac{\sum_{t=1}^{n-l} \hat{a}_t \hat{b}_{t+j}}{\left(\sum_{t=1}^{n} \hat{a}_t^2 \sum_{t=1}^{n} \hat{b}_t^2 \right)^{1/2}} \tag{3.19}$$

is calculated for $j = 0, \pm 1, \ldots, \pm n$, where $\hat{a}_t = x_t - \hat{\phi}_x x_{t-1}$ and $\hat{b}_t = y_t - \hat{\phi}_Y y_{t-1}$. The statistic $r_{\hat{a}\hat{b}}(k)$ is then an obvious choice for testing H_0. In fact, Jenkins and Watts (1968, p.340) show that when the univariate model for x_t and y_t is used to obtain estimates for ϕ_X, ϕ_Y, and ρ, the residual cross-correlation is the maximum likelihood estimate of ρ. Haugh (1976) has proved that under H_0, $r_{\hat{a}\hat{b}}(k)$ is asymptotically $N(0, \frac{1}{n})$; thus a test of asymptotic size α is obtained by rejecting H_0 whenever $\sqrt{n}|r_{\hat{a}\hat{b}}(k)| > Z(1 - \frac{\alpha}{2})$, where $Z(1 - \frac{\alpha}{2})$ denotes the $100(1 - \frac{\alpha}{2})\%$ quantile of the standard normal distribution. McLeod (1977) has shown that under H_1, $r_{\hat{a}\hat{b}}(k)$ is asymptotically $N(\rho, \frac{1}{n}(1 - \rho^2)^2)$. Thus, $r_{\hat{a}\hat{b}}(k)$ has the same large sample distribution as the ordinary sample correlation coefficient (Anderson, 1958, p.77). In fact, the test based on $r_{\hat{a}\hat{b}}(k)$ is asymptotically fully efficient by the following lemma.

Lemma 3.1 *Let the zero mean time series x_t and y_t satisfy*

$$\phi_X(B)X_t = a_t$$

and

$$\phi_Y(B)Y_t = b_t \ ,$$

where

$$\phi_X(B) = 1 - \phi_{X1}B \cdots - \phi_{Xp_X}B^{p_X}$$

and

$$\phi_Y(B) = 1 - \phi_{Y1}B \cdots - \phi_{Yp_Y}B^{p_Y}$$

such that all the roots of $\phi_X(B)$ and $\phi_Y(B)$ lie outside the unit circle, and $\rho_{ab}(l) = 0$ for all l except possibly at $l = k$. Then the test based on $r_{\hat{a}\hat{b}}(k)$ is asymptotically equivalent to a likelihood ratio test of H_0 against H_1.

Proof. Without loss of generality, assume $k = 0$.

Let the zero mean time series x_t and y_t satisfy, respectively

$$\begin{aligned} \phi_X(B)X_t &= a_t \\ \phi_Y(B)Y_t &= b_t \ , \end{aligned} \tag{3.20}$$

where

$$\phi_X(B) = 1 - \phi_{X1}B \cdots - \phi_{Xp_X}B^{p_X}$$

and

$$\phi_Y(B) = 1 - \phi_{Y1}B \cdots - \phi_{Yp_Y}B^{p_Y} ,$$

where all the roots of $\phi_X(B)$ and $\phi_Y(B)$ lie outside the unit circle. Under H_0 $(a_t, b_t)^{\mathrm{T}}$ is $N(\mathbf{0}, \mathbf{\Delta})$ distributed where $\mathbf{\Delta} = \begin{bmatrix} \sigma_a^2 & 0 \\ 0 & \sigma_b^2 \end{bmatrix}$ and under H_1 $(a_t, b_t)^{\mathrm{T}}$ is $N(\mathbf{0}, \mathbf{\Delta})$ distributed with $\mathbf{\Delta} = \begin{bmatrix} \sigma_a^2 & \rho \\ \rho & \sigma_b^2 \end{bmatrix}$. The likelihood ratio statistic of testing H_0 against H_1 is given by

$$\lambda \propto \frac{\hat{\sigma}_a^{-\frac{n}{2}} \hat{\sigma}_b^{-\frac{n}{2}} \exp -\left(\sum \hat{a}_t^2 / 2\hat{\sigma}_a^2\right) \exp -\left(\sum \hat{b}_t^2 / 2\hat{\sigma}_b^2\right)}{|\hat{\mathbf{\Delta}}|^{-\frac{n}{2}} \exp -\left(\sum \hat{e}_t^{\mathrm{T}} \hat{\mathbf{\Delta}}^{-1} \hat{e}_t / 2\right)}$$

where $\hat{a}_t(\hat{b}_t)$ and $\hat{\sigma}_a^2(\hat{\sigma}_b^2)$ are the residuals and the maximum likelihood estimates under H_0 of $a_t(b_t)$ and $\sigma_a^2(\sigma_b^2)$, respectively; $\hat{\mathbf{\Delta}}$ and $\hat{\mathbf{e}}_t^{\mathrm{T}} = (\bar{a}_t, \bar{b}_t)$ are the estimates of $\mathbf{\Delta}$ and the residuals of fitting the bivariate series $(X_t, Y_t)^{\mathrm{T}}$ under H_1 using maximum likelihood procedure. It is well known that $\hat{\sigma}_a^2(\hat{\sigma}_b^2) = \sum \hat{a}_t^2 / n \left(\sum \hat{b}_t^2 / n\right)$ and $\hat{\mathbf{\Delta}} = \sum \hat{\mathbf{e}}_t \hat{\mathbf{e}}_t^{\mathrm{T}} / n$ and hence

$$\lambda \propto \frac{\hat{\sigma}_a^{-\frac{n}{2}} \hat{\sigma}_b^{-\frac{n}{2}}}{|\hat{\mathbf{\Delta}}|^{-\frac{n}{2}}}$$

Now, conditional on the first $p = \max(p_X, p_Y)$ observations, the maximum likelihood estimator of the bivariate model for (X_t, Y_t), is up to probability order $1/\sqrt{n}$ given by the univariate Yule-Walker equations of X_t and Y_t on the diagonal and 0 elsewhere. Hence, $\hat{\mathbf{e}}_t^{\mathrm{T}} = (\bar{a}_t, \bar{b}_t)$ can be considered as asymptotically the same as $(\hat{a}_t, \hat{b}_t)$ and thus λ is asymptotically proportional to

$$\begin{aligned} & \frac{\sum(\hat{a}_t^2/n) \sum(\hat{b}_t^2/n) - \left(\frac{\sum \hat{a}_t \hat{b}_t}{n}\right)^2}{\sum(\hat{a}_t^2/n) \sum(\hat{b}_t^2/n)} \\ = \quad & 1 - r_{\hat{a}\hat{b}}(0)^2 . \end{aligned}$$

The lemma follows. □

It is instructive to compare the above test with one based on the sample cross-correlations

$$r_{xy}(l) = \frac{\sum_{1}^{n-l} x_t y_{t+l}}{\sqrt{\left(\sum_{1}^{n} x_t^2\right)\left(\sum_{1}^{n} y_t^2\right)}}$$

The asymptotic variances of $r_{xy}(l)$ can be computed from a formula of

Bartlett (1966, p.349) and it can be seen that, in general, these variances depend on the unknown parameters ϕ_X and ϕ_Y. However, if $\phi_Y = 0$, then as pointed out by Bartlett (1935), the large sample distribution of $r_{xy}(k)$ under H_0 is $N\big(0, \frac{1}{n}\big)$ and hence a test of asymptotic size α can be defined by rejecting H_0 whenever $\sqrt{n}\big|r_{xy}(k)\big| > Z\big(1-\frac{\alpha}{2}\big)$. This situation may arise when one of the series is completely uncorrelated, as in the example of Bartlett (1935, p.542) of the relationship between a climatic index and a mortality index. Denote the theoretical autocorrelations of X_t by $\rho_{XX}(k)$, $k = 0, \pm 1, \ldots$. Similarly denote the theoretical cross-correlation between X_t and Y_t by $\rho_{XY}(k)$, $k = 0, \pm 1, \ldots$.

Lemma 3.2 *Under H_1, when $\phi_Y = 0$, $k = 0$, $r_{xy}(k)$ is asymptotically*

$$N\Big(\rho\sqrt{(1-\phi_X^2)}, \frac{1}{n}\left[1-\rho^2\left(1+\left(1-\phi_X^2\right)\left(1-\rho^2\right)\right)\right]\Big)$$

Proof. From Bartlett's formula

$$\begin{aligned}
n \cdot \operatorname{var}\big(r_{xy}(0)\big) &= \sum_{i=-\infty}^{\infty} \rho_{XX}(i)\rho_{YY}(i) + \rho_{YX}(i)\rho_{XY}(i) \\
&\quad + \rho_{XY}^2(0)\left[\rho_{XY}^2(i) + \frac{1}{2}\rho_{XX}^2(i) + \frac{1}{2}\rho_{YY}^2(i)\right] \\
&\quad - 2\rho_{XY}^2(0)[\rho_{XX}(i)\rho_{YX}(i) + \rho_{YX}(i)\rho_{YY}(i)] \\
&= 1 + \rho^2(1-\phi_X^2) + \rho^2(1-\phi_X^2)\left[\rho^2 + \frac{1}{1-\phi_X^2}\right] \\
&\quad - 2\rho\sqrt{1-\phi_X^2} \cdot \left(\rho\sqrt{1-\phi_X^2} + \frac{\rho}{\sqrt{1-\phi_X^2}}\right) \\
&= 1 - \rho^2\left[1 + \left(1-\phi_X^2\right)\left(1-\rho^2\right)\right] .
\end{aligned}$$

The lemma thus follows. □

It can be seen from above that $r_{xy}(0)$ has smaller mean and larger variance than $r_{\hat{a}\hat{b}}(0)$ (provided $\phi_x \neq 0$). If the "power" of a test is defined to be the large sample approximation to the probability that it will reject H_0, it follows that the test based on $r_{\hat{a}\hat{b}}(0)$ is "uniformly" more powerful than the test based on $r_{xy}(0)$. Figure 3.1 is a plot of the corresponding powers of the two tests when $n = 200$ and $\phi_X = 0.8$. As can be seen, the difference in "power" between these tests can be considerable. The results of a simulation experiment on the empirical power of these tests under the conditions of Lemma 2 are given in Table 3.3. There are 1000 replications for each combination of values of ϕ_X and ρ used and the

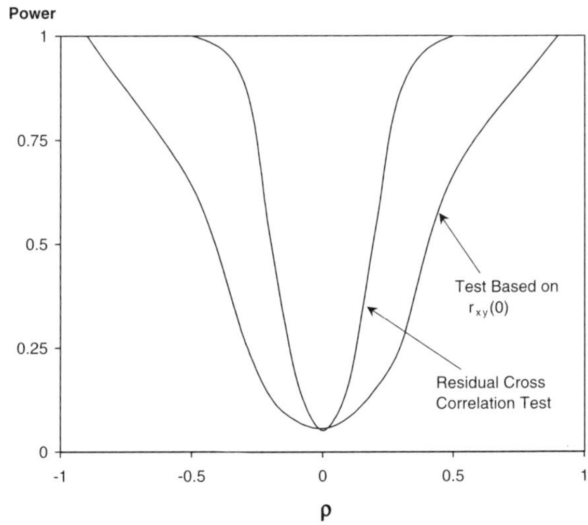

Figure 3.1 *Power when* $n = 200$ *and* $\phi_x = 0.9$, $\alpha = .05$.

number of times that H_0 is rejected, at $\alpha = .05$, is recorded. The lengths of all the series are equal to 100.

Table 3.3 *Empirical comparison of* $r_{xy}(0)$ *and* $r_{\hat{a}\hat{b}}(0)$

ϕ_x	ρ	No. of rejections at $\alpha = 0.5$ of H_0 using $r_{xy}(0)$	No. of rejections at $\alpha = .05$ of H_0 using $r_{\hat{a}\hat{b}}$
$-.9$	$-.9$	1000	1000
	$-.5$	644	1000
	$-.3$	262	875
	$-.2$	150	533
	$-.1$	63	157
	0	46	59
	.1	81	166
	.2	149	507
	.3	257	872
	.5	621	1000
	.9	1000	1000

(Cont.)

Table 3.3 *Empirical comparison of* $r_{xy}(0)$ *and* $r_{\hat{a}\hat{b}}(0)$

ϕ_x	ρ	No. of rejections at $\alpha = 0.5$ of H_0 using $r_{xy}(0)$	No. of rejections at $\alpha = .05$ of H_0 using $r_{\hat{a}\hat{b}}$
−.5	−.9	1000	1000
	−.5	998	1000
	−.3	774	880
	−.2	449	541
	−.1	138	157
	0	55	49
	.1	132	187
	.2	422	505
	.3	769	873
	.5	997	1000
	.9	1000	1000
−.1	−.9	1000	1000
	−.5	999	999
	−.3	861	859
	−.2	521	520
	−.1	173	166
	0	40	38
	.1	167	164
	.2	538	538
	.3	861	859
	.5	1000	1000
	.9	1000	1000
0.0	−.9	1000	1000
	−.5	1000	1000
	−.3	894	889
	−.2	517	512
	−.1	188	189
	0	37	41
	.1	167	163
	.2	535	535
	.3	888	877
	.5	1000	1000
	.9	1000	1000

(Cont.)

Table 3.3 *(Continued)*

ϕ_x	ρ	No. of rejections at $\alpha = 0.5$ of H_0 using $r_{xy}(0)$	No. of rejections at $\alpha = .05$ of H_0 using $r_{\hat{a}\hat{b}}$
.1	−.9	1000	1000
	−.5	1000	1000
	−.3	883	881
	−.2	559	553
	−.1	170	174
	0	38	38
	.1	156	157
	.2	527	534
	.3	862	864
	.5	1000	1000
	.9	1000	1000
.5	−.9	1000	1000
	−.5	998	1000
	−.3	776	890
	−.2	413	514
	−.1	135	171
	0	45	59
	.1	145	180
	.2	455	541
	.3	758	859
	.5	999	1000
	.9	1000	1000
.9	−.9	1000	1000
	−.5	640	1000
	−.3	271	887
	−.2	132	512
	−.1	74	175
	0	56	51
	.1	83	164
	.2	149	526
	.3	256	872
	.5	666	999
	.9	1000	1000

It can be seen that except when ϕ_X or ρ is nearly zero, the test based on $r_{\hat{a}\hat{b}}(0)$ is far more sensitive in rejecting correctly the null hypothesis and in those cases where $r_{xy}(0)$ appears to be better the differences are not significant. Sims (1977) raised the question of bias for the test based on $r_{\hat{a}\hat{b}}(k)$. It may be concluded on the basis of the above results, that at least in the case of instantaneous causality, the univariate approach to residual cross-correlation should be recommended.

3.3 Transfer function noise (TFN) modeling

For simplicity we consider $l = 2$ in (3.1) and $\mathbf{\Theta}(B) = \mathbf{1}_l$. Suppose also that the Φ_is are of the lower diagonal form $E(\boldsymbol{a}_t \, \boldsymbol{a}_t^{\mathrm{T}}) = \mathbf{\Delta}$ is diagonal. Then (3.1) can be written

$$\begin{aligned} X_t - \phi_{1,11}\, X_{t-1} - \cdots - \phi_{p,11}\, X_{t-1} &= a_{1t} \\ Y_t - \phi_{1,21}\, X_{t-1} - \cdots - \phi_{p,21}\, X_{t-1} & \\ - \;\phi_{1,22} Y_{t-1} - \cdots - \phi_{p,22}\, Y_{t-1} &= a_{2t} \;, \end{aligned}$$

where $\phi_{l,ij}$ is the (i,j)th element of Φ_l. The equation for Y_t above can be viewed as a special case of the TFN model with one time series X_t as input and Y_t as output. More generally, assuming that (X_t, Y_t) is stationary a TFN model for Y_t as output and X_t as input with no feedback is given by

$$Y_t - \mu_Y = \frac{\omega(B)}{\delta(B)}(X_{t-b} - \mu_X) + \frac{\theta(B)}{\phi(B)}\, a_t \tag{3.21}$$

where b is the delay, $\omega(B)$, $\delta(B)$, $\theta(B)$, $\phi(B)$ are polynomials in B with orders s, r, q, and p, respectively, $\mu_Y = E(Y_t)$, $\mu_X = E(X_t)$, and a_t is white noise. The coefficient of B^0 in $\delta(B)$ is one while that of $\omega(B)$ is an (unknown) constant. Note that $\nu(B) = \omega(B)/\delta(B) = \nu_0 + \nu_1 B + \nu_2 B^2 + \cdots$ is called the transfer function and ν_0, ν_1, $\nu_2, \ldots$ are called the impulse responses. The noise series N_t is given by $\theta(B)/\phi(B)a_t$. It is assumed that $\{a_t\}$ and $\{X_t\}$ are independent.

By following the procedure of Box and Jenkins (1976) the TFN model (3.21) can be constructed according to the following steps which are based on prewhitening the input X_t.

(i) Assume that X_t and Y_t satisfy the ARMA model specifications (3.12a) and (3.12b) respectively. Determine the most appropriate ARMA model to fit to the x_t series by utilizing the three stages of model construction (Box and Jenkins, 1976). At the estimation stage, estimates are obtained for the ARMA model parameters and also the innovation series $\hat{u}_t$.

(ii) Using the ARMA filter, $\hat{\theta}_X(B)/\hat{\phi}_X(B)$, from step (i), transform the y_t series using

$$\hat{\beta}_t = \left|\frac{\hat{\theta}_X(B)}{\hat{\phi}_X(B)}\right|^{-1} y_t \tag{3.22}$$

where the $\hat{\beta}_t$ sequence is usually not white noise.

(iii) Calculate the residual cross-correlation function (CCF) $r_{\hat{u}\hat{\beta}}(k)$ for the $\hat{u}_t$ and $\hat{\beta}_t$ series.

(iv) Based upon the behavior of the residual CCF from step (iii), identify the parameters required in the transfer function, $\nu(B)$, in (3.21). As shown by Box and Jenkins (1976, p.380), the theoretical CCF $\rho_{u\beta}(k)$ between the prewhitened input, u_t, and the correspondingly transformed output, β_t, is related to the impulse response function ν_k by the expression

$$\nu_k = \rho_{u\beta}(k)\frac{\sigma_\beta}{\sigma_a}$$

where σ_β and σ_a are the standard deviations of β_t and a_t respectively. Hence moment estimates of ν_k can be obtained using this relation.

(v) Given initial moment estimates for the parameters in $\nu(B)$, estimate the noise series from (3.21) by using

$$\hat{N}_t = (y_t - \bar{y}) - \hat{\nu}(B)(x_t - \bar{x})$$

where $\bar{y}$ and $\bar{x}$ are the sample means for μ_Y and μ_X, respectively. The forms of $\omega(B)$ and $\delta(B)$ can also be identified tentatively using the patterns of $\hat{\nu}_i$ as suggested by Box and Jenkins (1976). By examining the sample autocorrelation function (ACF) and the sample partial autocorrelation function of $\hat{N}_t$, identify the ARMA model needed to fit to the noise series. The entire transfer function-noise model has now been tentatively identified. Maximum likelihood estimation can then be applied to estimate the model parameters simultaneously.

Haugh and Box (1977) proposed an alternative approach where both X_t and Y_t are prewhitened by an appropriate ARMA filter, respectively. The impulse response weights are then estimated by the CCF of the respective residuals. An advantage of the Haugh and Box method is that the residual CCF results that are employed for detecting causal relationships are also used for model identification.

The innovation sequence, a_t, is often assumed to be independently distributed and a recommended procedure for checking the whiteness assumption is to examine a plot of the residual ACF along with confidence limits. Denote the residual ACF by $r_{\hat{a}\hat{a}}(k)$. Since $r_{\hat{a}\hat{a}}(k)$ is symmetric about lag zero, the residual ACF is plotted against lags for $k = 1$ to $k \simeq n/4$ or $n/5$ and the method of McLeod (1978) can be employed to calculate confidence limits. If the residuals are correlated, this suggests some type of model inadequacy. To determine the source of the error in the model, the CCF $r_{\hat{u}\hat{a}}(k)$ for the $\hat{u}_t$ and $\hat{a}_t$ sequences can be studied. Because the X_t and a_t series are assumed to be independent of one another, the estimated values of $r_{\hat{u}\hat{a}}(k)$ should not be significantly different from zero. Note that the 95% confidence limits for the CCF are about plus and minus two times $n^{-1/2}$ when the sample size is large. When a plot of $r_{\hat{u}\hat{a}}(k)$ indicates whiteness while significant correlations are present in $r_{\hat{a}\hat{a}}(k)$, the model inadequacy is probably in the noise term, $\hat{N}_t$. As in the ARMA case, the form of the residual ACF for the $\hat{a}_t$ series could suggest appropriate modifications to the noise structure. However, if both $r_{\hat{a}\hat{a}}(k)$ and $r_{\hat{u}\hat{a}}(k)$ possess one or more significant values, this could mean that the transfer function is incorrect and the noise term may or may not be suitable. By a result of Pierce (1972),

$$S_m = n \cdot \sum_{k=0}^{m} r^2_{\hat{u}\hat{a}}(k)$$

is approximately χ^2_{m-r-s} distributed and therefore S_m may also be used as a model diagnostic statistic.

When feedback is indicated by significant values of $r_{\hat{u}\hat{a}}(k)$ at negative lags, a multivariate ARMA model should be considered rather than a transfer function-noise model. Whenever problems arise in the model building process, suitable model modifications can often be made from information at the diagnostic checking and identification stages.

The a_t sequence is often assumed to possess constant variance (homoscedasticity) and follow a normal distribution. Tests are available for checking the homoscedastic and normality suppositions (see, for example, Hipel et al. (1977), McLeod et al. (1977) and Chapter 6 of this book), and in practice it has been found that suitable Box-Cox transformations of the Y_t and/or the X_t series may correct heteroscedasticity and nonnormality in the residuals. Nelson and Granger (1979), however, suggest that the Box-Cox transformation does not consistently produce better forecasts. The Box-Cox transformation for the Y_t series is given as

$$Z_t = \begin{cases} \lambda^{-1}\left[\left(Y_t + c\right)^{\lambda} - 1\right] & \lambda \neq 0 \\ \ln\left(Y_t + c\right) & \lambda = 0 \end{cases}$$

where the constant c is usually assigned a magnitude which is just large enough to make all the entries in the Y_t series positive. See Hipel,

McLeod, and Li (1985) and Hipel and McLeod (1994) for more details. Atkinson (1986) discussed diagnostic tests for transformations under the regression context. His methods can be extended to the time series context.

CHAPTER 4

Robust modeling and diagnostic checking

4.1 A robust portmanteau test

As in other fields of statistics, the presence of outliers can present serious problems in time series modeling. There are two types of outliers in time series. We have the innovation outliers (IO) if the noise process a_t has a heavy tailed distribution compared with the normal distribution. This type of outlier is less problematic if a_t has finite fourth order moment. It can be shown that in this situation the conditional least squares estimators obtained by minimizing (2.3b) will still be consistent with the same covariance matrix given by the inverse of $\mathbf{I}$ in (2.5). Another more serious type of outliers is known as the additive outliers (AO). Additive outliers are present if instead of X_t we observe $z_t = X_t + W_t$, where $\{X_t\}$ follow the ARMA time series (2.1), and W_t is a contaminating process with $P(W_t \neq 0) = C$ for some C with $0 \leq C \leq 1$. The presence of W_t masks the original autocorrelation structure of X_t and hence causes greater problems in the modeling of X_t. Note that in many applications W_t is assumed to be independent, identically distributed, and sometimes assumes a fixed value δ.

As an illustration of the effect of additive outliers we consider the installation of residential telephone extensions series (RESEX) from Martin, Samarov, and Vandaele (1983). The data set is also listed in Rousseeuw and Leroy (1987). Figure 4.1 shows the time series plot of the original series and Figure 4.2 gives the sample autocorrelations and partial autcorrelations of the seasonally differenced series using the software *ITSM* in Brockwell and Davis (1996). It can be seen from Figure 4.1 that the observations at $t = 83$ and 84 are somewhat larger than the rest of the series and may be regarded as outliers. From Figure 4.2 the series can be identified as an AR(1) process because the partial autocorrelation has a cut-off after lag 1. The two observations were then replaced by observations from the same months in the previous year (1971). Figure 4.3 gives the time series plot of the outlier adjusted series and Figure 4.4 gives the sample autocorrelations and the partial autocorrelation. It can be seen that the dependence structure of the series is now much stronger and the partial autcorrelations suggested an AR(2) model instead. Fox (1972) gave a comprehensive account of outliers in time series. Whether outliers should be removed or how they should be removed are controversial issues. An alternative route is to protect the statistical procedure that one is using from the effects of outliers. Here we will concentrate on this latter approach by emphasizing robust time series estimation and robust goodness-of-fit tests. When X_t follows an autoregressive model of order p, Martin (1980) proposed generalized M (GM)-estimators for the autoregressive parameters ϕ_i and the scale parameter s of a_t. However, the asymptotic covariance matrix of the GM-estimates does not have a closed form in the general situation under AO. Martin (1982), Lee and Martin (1986), and Masarotto (1987) gave some further results of GM estimates. Bustos and Yohai (1986) proposed an alternative set of robust estimates by robustifying the conditional least squares estimation equations. Lo and Li (1990) considered robust Yule-Walker estimates and least squares estimates using robustified autocorrelation and covariance matrices.

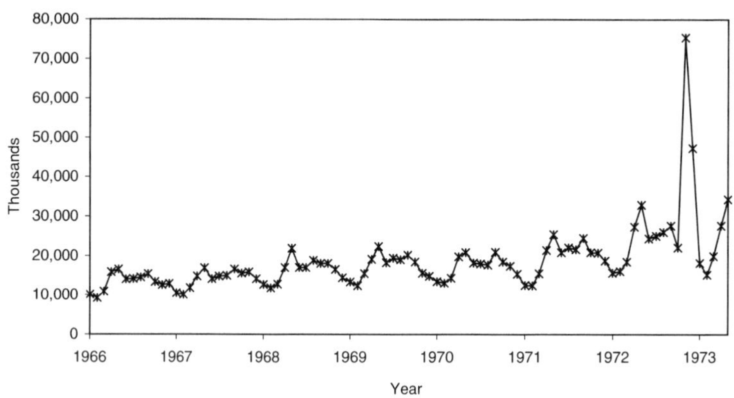

Figure 4.1 *Time series plot of the RESEX series (original data)*

Detection of outliers and estimation based on the intervention analysis approach (Box and Tiao, 1975) have been considered by Tsay (1988), Chang, Tiao and Chen (1988), and Abraham and Chuang (1989). In these papers likelihood ratio type tests have been developed to detect outliers and identify their types. If the positions of outliers are unknown, their impact can usually be modeled using a dummy variable approach such as intervention analysis. Outlier detection using the influence func-

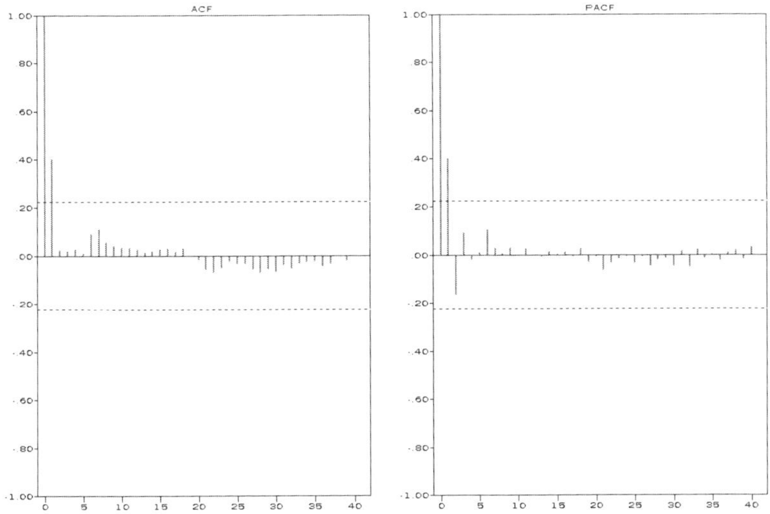

Figure 4.2 *Sample autocorrelations and partial autocorrelations of the seasonally differenced RESEX series*

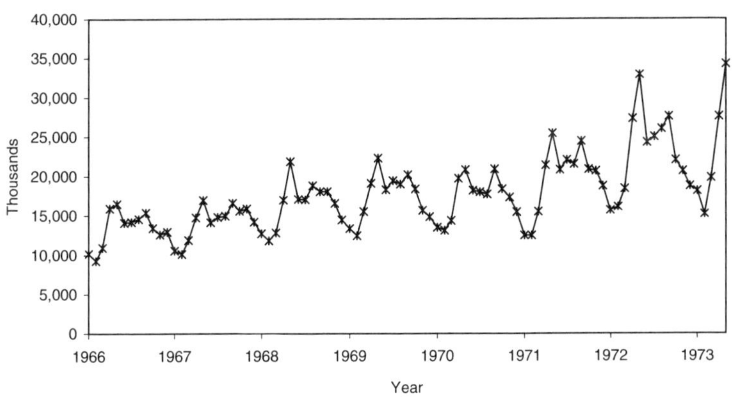

Figure 4.3 *Time series plot of the adjusted RESEX series*

tion was considered by Chernick, Downing and Pike (1982), and Bruce and Martin (1989). The field of outlier detection in time series is immense and it is beyond the scope of this monograph to give a detailed

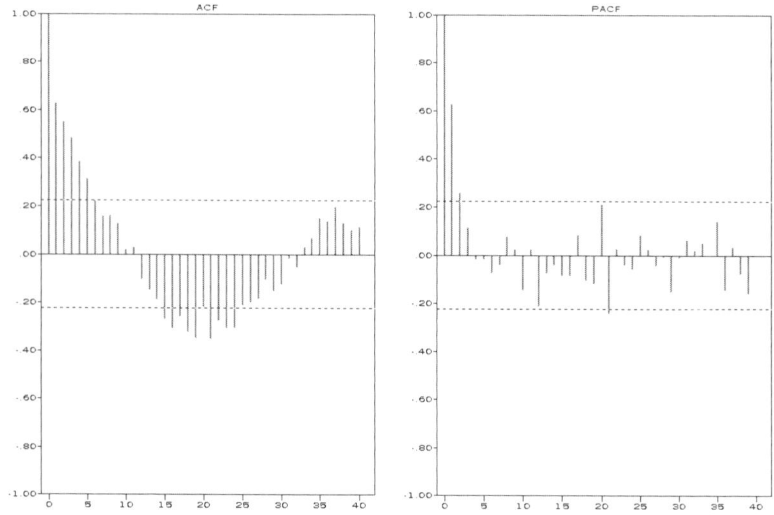

Figure 4.4 *Sample autocorrelations and partial autocorrelations of the seasonally differenced RESEX series with adjustments for outliers*

account of the topic. Readers are referred to the above papers for more details.

Without loss of generality let the mean $\mu = 0$, otherwise we can always center the time series with a robust estimator of the mean. It is also assumed that a_t is symmetrically distributed about zero. Let the vector of AR and MA parameters be $\boldsymbol{\beta}^{\mathrm{T}} = (\phi_1, \ldots, \phi_p, \theta_1, \ldots, \theta_q)$. Given $\{z_t\}$ for $t = 1, \ldots, n$, the estimating equations of the least squares or the conditional likelihood estimator of $\boldsymbol{\beta}$ can be written as

$$\sum_{h=0}^{n-j-p-1} \phi'_h r_{h+j} = 0 \ , \qquad \sum_{h=0}^{n-j-p-1} \theta'_h r_{h+j} = 0 \ , \tag{4.1}$$

where $r_j = \sum a_t a_{t-j}$, $a_1 = \cdots = a_p = 0$; $\phi^{-1}(B) = \sum \phi'_h B^h$; and $\theta^{-1}(B) = \sum \theta'_h B^h$. By robustifying r_j, Bustos and Yohai (1986) suggested the so-called residual autocovariance (RA) estimator. The robustification of r_j is done by defining

$$\gamma_j = \sum_{t=p+1+j}^{n} \eta\big(a_t/\hat{\sigma}, a_{t-j}/\hat{\sigma}\big) \tag{4.2}$$

where "/" denotes the division sign, $\hat{\sigma}$ is a robust scale estimate, and η

is an odd function in each variable. The function η may be chosen to be either of the Mallows type: $\eta(u, v) = \psi(u)\psi(v)$; or of the Hampel type: $\eta(u, v) = \psi(u, v)$, where ψ is a continuous odd function. For example, ψ may be of the Huber family:

$$\psi_{H,c}(u) = \text{sgn}(u)\min(|u|, c) \ ,$$

or the bisquares family:

$$\psi_{B,c}(u) = u(1 - u^2/c^2)^2 \quad (0 \le |u| \le c) \ .$$

By choosing $\eta(u, v) = \psi(u)v$ and $\eta(u, v) = uv$ the residual autocovariance estimator gives Huber's M-estimator and the conditional likelihood estimator, respectively. For the Mallows type η an iteratively weighted least squares scheme for estimating $\boldsymbol{\beta}$ is possible. A nonlinear optimization routine would have to be employed in general. Since η is an odd function, $E\{\eta(a_t/\sigma, a_{t-i}/\sigma)\} = 0$, $i \neq 0$, where the expectation is taken with respect to the distribution of a_t. Bustos and Yohai (1986) showed that $\sqrt{n}(\hat{\boldsymbol{\beta}} - \boldsymbol{\beta})$ is asymptotically normally distributed with mean zero and covariance matrix $v\mathbf{I}^{-1}$, where $\mathbf{I}^{-1}$ is the covariance matrix of the usual conditional likelihood estimates (see (2.5)) and $v = a\sigma^2/b^2$, where

$$a = E\{\eta^2(a_t/\sigma, a_{t-1}/\sigma)\} \ , \quad b = E\{\eta_1(a_t/\sigma, a_{t-1}/\sigma)a_{t-1}\} \qquad (4.3)$$

with $\eta_1(u, v) = \partial\eta(u, v)/\partial u$. Bustos and Yohai (1986) demonstrated that the RA estimates have good robustness properties, in particular, against AO's.

Li (1988) derived a robustified portmanteau goodness-of-fit test for ARMA time series models estimated using the RA estimators which was based on the asymptotic distribution of a robust residual autocorrelation function resulting from the RA estimates.

Denote by $\hat{a}_t$ the residuals obtained when $\boldsymbol{\beta}$ is estimated by the method discussed in (4.1). Let

$$\hat{\gamma}_j = \sum_{t=p+j+1}^{n} \eta(\hat{a}_t/\hat{\sigma}, \hat{a}_{t-j}/\hat{\sigma})/n \ , \quad R_j = \sum_{t=p+j+1}^{n} \eta(a_t/\sigma, a_{t-j}/\sigma)/n \ ,$$

where $\hat{\sigma}$ is as before a robust scale estimator. Define $\hat{\boldsymbol{\gamma}}^{\mathrm{T}} = (\hat{\gamma}_1, \ldots, \hat{\gamma}_m)$ for some $m > 0$. Similarly define $\boldsymbol{R}^{\mathrm{T}}$. Suppose that all relevant expectations exist. Bustos, Fraiman, and Yohai (1984) obtained the result that $\hat{\boldsymbol{\beta}}$ and $\hat{\sigma}$ are asymptotically uncorrelated and $\hat{\sigma}$ has variance of order n^{-1}. Since $\eta(u, v)$ is odd in each variable it can be seen that

$$E\{\eta(a_t/\sigma, a_{t-j}/\sigma)\eta(a_{t'}/\sigma, a_{t'-k}/\sigma)\} = 0$$

if $t \neq t'$ or $j \neq k$. The following lemmas can then be obtained as in Li and McLeod (1981) and McLeod (1978). Note that the random vector $\sqrt{n}\boldsymbol{R}$

is asymptotically normally distributed with mean zero and covariance matrix $a\mathbf{1}_m$, where $\mathbf{1}_m$ is the $m \times m$ identity matrix and a is defined in (4.3).

Lemma 4.1 *For large n, $\hat{\gamma} = \boldsymbol{R} - b\sigma^{-1}\mathbf{X}(\hat{\boldsymbol{\beta}} - \boldsymbol{\beta}) + O_p(n^{-1})$, where $\mathbf{X} = (\phi'_{i-j}|\theta'_{i-j})$ is an $m \times (p+q)$ matrix defined in Chapter 2.*

Lemma 4.2 *The asymptotic cross-covariance of $\sqrt{n}(\hat{\boldsymbol{\beta}}-\boldsymbol{\beta})$ and $\sqrt{n}(\boldsymbol{R})$ is $(a\sigma/b) \cdot (\mathbf{I}^{-1}\mathbf{X}^{\mathrm{T}})$.*

The following theorem follows by combining the above lemmas.

Theorem 4.1 *The asymptotic distribution of $\sqrt{n}\hat{\gamma}$ is Gaussian with mean zero and covariance matrix $a(\mathbf{1}_m - \mathbf{X}\mathbf{I}^{-1}\mathbf{X}^{\mathrm{T}})$.*

It follows at once from the classical result in Chapter 2 that, if $n \gg m \gg 0$, then $\sqrt{(n/a)}\hat{\gamma}$ has an asymptotic covariance matrix that is idempotent of rank $m-p-q$. Hence, the statistic

$$Q_m = a^{-1}n\sum_{k=1}^{m}\hat{\gamma}_k^2 \tag{4.4}$$

is asymptotically distributed as chi-squared with degrees of freedom $m-p-q$. Note that, as in the classical Gaussian situation, $E(Q_m) \neq m-p-q$ for moderate values of m and n. Therefore it is natural to adjust the statistic by either the Li and McLeod (1981) approach or by the factor $(n+2)/(n-k)$ as in Ljung and Box (1978). Note that it can also be shown that there is a $1-1$ correspondence between $\hat{\gamma}_i$ and the estimating equations

$$L_{p+i} = n\sum_{h=0}^{n-2p-i-1}\hat{\phi}'_h\hat{\gamma}_{h+p+i} \quad (1 \leq i \leq k)\ , \tag{4.5}$$

where $L_j = 0$ for $1 \leq j \leq p$, and the quantities in (4.5) are evaluated using RA estimates from an autoregressive model of order p. See Li (1988). This gives a robustified result of Newbold (1980) where it is shown that the Lagrange multiplier test of AR(p) *vs* AR($p+k$) is equivalent to a test based on the first k residual autocorrelations.

In Li (1988) the robustness of the proposed statistics in the presence of outliers has been studied by simulation. The robustness of the upper 10th and 5th percentiles of the $\tilde{Q}_{10}$ and Q_{10} statistics were investigated for a contaminated autoregressive process of order one; see Table 4.1. Here

$$\tilde{Q}_m = a^{-1}n^2\sum_{k=1}^{m}\hat{\gamma}_k^2/(n-k)\ .$$

Table 4.1 *Empirical mean, variance, and upper 10th and 5th percentiles of* Q_m *and* $\tilde{Q}_m$, $m = 10$ *(Li, 1988). ©1988 Biometrika Trust, reproduced with the permission of Oxford University Press*

	(a) No outliers							
	Q_m				$\tilde{Q}_m$			
ϕ_1	Mean	var	10%	5%	Mean	var	10%	5%
0.5	9.20	18.69	15.00	17.49	9.20	18.94	15.05	17.12
−0.5	9.11	19.08	14.65	17.91	9.14	19.03	14.85	17.89
0.8	9.49	20.55	15.41	17.58	9.53	20.80	15.54	17.90
−0.8	9.29	20.16	15.50	17.46	9.32	20.28	15.52	17.53
	(a) With additive outliers							
	Q_m				$\tilde{Q}_m$			
ϕ_1	Mean	var	10%	5%	Mean	var	10%	5%
0.5	3.30	2.71	5.47	6.58	8.09	13.29	12.90	14.82
−0.5	3.37	3.37	5.87	6.72	8.20	15.43	13.41	15.62
0.8	6.39	15.11	11.86	13.84	9.89	20.25	15.88	18.46
−0.8	6.48	13.16	11.81	13.33	9.92	19.96	15.70	17.96

There were 1000 replications, each of length 100, for each parameter value. The a_t's were from an $N(0,1)$ population generated by the IMSL subroutine GGNPM. The outliers were fixed at $t = 11, 33, 49, 76$, and 90 and had values 10, −10, 10, −10, 10, respectively. Mallows type $\eta(u,v)$ and Huber's ψ function with tuning constant $c = 2{\cdot}52$ were used for the residual autocovariance estimates. The scale parameter was estimated by the median of $(|a_{p+1}|, \ldots, |a_n|)/(0{\cdot}6745)$.

Table 4.1 shows that where there were no outliers the $\tilde{Q}_m$ statistic mimics the Q_m statistics closely in all aspects considered. However, if outliers were present, the Q_m statistic differed significantly from that of a chi-squared random variable with nine degrees of freedom. On the other hand, the distribution of $\tilde{Q}_m$ appeared reasonably approximated by the asymptotic theory. Li (1988) further demonstrated that, with outliers present, the power of $\tilde{Q}_m$ is much better than that of Q_m. The results are not repeated here.

4.2 A robust residual cross-correlation test

Based on the univariate RA estimates of the previous section it is natural to construct a robust residual cross-correlation test for lagged relations in time series. This result has applications in robust Granger causality tests and robust transfer function noise modeling which were discussed in Chapter 3. We follow the approach of McLeod (1979) and Li and Hui (1994). Following the notation of Li and Hui (1994) denote by $\{X_{h,t}\}$, $h = 1, 2$, the two time series under consideration. It is assumed that they satisfy the autoregressive moving average processes,

$$\phi_h(B)X_{h,t} = \theta_h(B)a_{h,t}, \quad h = 1, 2, \quad t = 1, 2, \dots , \tag{4.6}$$

where

$$\phi_h(B) = \sum_{i=0}^{p_h} \phi_{h,i} B^i \ , \quad \theta_h(B) = \sum_{i=0}^{q_h} \theta_{h,i} B^i \ , \quad \phi_{h,0} = \theta_{h,0} = 1 \ ,$$

B is the backward shift operator, and all the roots of $\phi_h(B)$ and $\theta_h(B)$ are outside the unit circle so that $\{X_{h,t}\}$, $h = 1, 2$, are stationary and invertible. For simplicity we assume that $E(X_{h,t}) = 0$ and $\theta_h(B) = 1$. For each h, $h = 1, 2$, the innovation series $\{a_{h,t}\}$ are assumed to be independent variates with mean zero and variance σ_h^2. However, a_{1t} and $a_{2t'}$ could be correlated. As in §4.1 we will assume that $\{a_{h,t}\}$ are symmetric about zero. Let $\phi_h^{-1}(B) = \sum \phi'_{h,i} B^i$. Let $\boldsymbol{\phi}_h = (\phi_{h1}, \dots, \phi_{hp_h})^{\mathrm{T}}$. Suppose the length of the realizations is n. As in §4.1 a robust residual autocovariance (RA) estimate of (4.6) is obtained by solving the system of estimating equations

$$L_{hj} = \sum_{i=0}^{n-j-p_h-1} \phi'_{h,i} \gamma_{h,i+j} = 0 \ , \quad j = 1, \dots, p_h \ , \quad h = 1, 2,$$

where

$$\gamma_{h,j} = \sum_{p_h+1+j}^{n} \eta(a_{h,t}/\sigma_h, a_{h,t-j}/\sigma_h) \ ,$$

with $\eta(u, v) = \psi(u)\psi(v)$ or $\psi(uv)$, where ψ is a continuous odd function in each variable; "/" denotes division. The scale parameters σ_h can be estimated jointly, for example, by $\text{med}\{|\hat{a}_{h,i}|\}/0.6745$ (Bustos and Yohai, 1986), where $\text{med}(\cdot)$ denotes the median and $|\cdot|$ denotes absolute values. Denote estimates of $\boldsymbol{\phi}_h$ by $\hat{\boldsymbol{\phi}}_h$ and the corresponding residual by $\hat{a}_{h,t}$. From the discussions before (4.3) $\sqrt{n}(\hat{\boldsymbol{\phi}}_h - \boldsymbol{\phi}_h)$ is asymptotically normally distributed with mean zero and covariance matrix $V_h = (\alpha_h \sigma_h^2/\beta_h^2)\mathbf{I}_h^{-1}$, where $\mathbf{I}_h^{-1}$ is the covariance matrix of the Gaussian likelihood estimates, $\beta_h = E[\eta_1(a_{h,t}/\sigma_h, a_{h,t-1}/\sigma_h)a_{h,t-1}]$ with $\eta_1(u, v) = \partial\eta(u, v)/\partial u$ and $\alpha_h = E[\eta^2(a_{h,t}/\sigma_h, a_{h,t-1}/\sigma_h)]$. Let the robustified lag l innovation cross-correlation be

$$\gamma_{a_1 a_2}(l) = n^{-1} \sum_{t=1}^{n-l} \eta(a_{1t}/\sigma_1, a_{2t+l}/\sigma_2) \ . \tag{4.7}$$

Similarly define the robustified residual cross-correlations $\hat{\gamma}_{a_1a_2}(l)$ by replacing $a_{h,t}$ with $\hat{a}_{h,t}$ in the above expression. Let

$$\boldsymbol{\gamma} = (\gamma_{a_1a_2}(-1), \ldots, \gamma_{a_1a_2}(-M),\ \gamma_{a_1a_2}(0), \ldots, \gamma_{a_1a_2}(1), \ldots, \gamma_{a_1a_2}(M))^{\mathrm{T}} .$$

Let

$$\boldsymbol{\rho} = (\rho_{a_1a_2}(-1), \ldots, \rho_{a_1a_2}(-M),\ \rho_{a_1a_2}(0), \rho_{a_1a_2}(1), \ldots, \rho_{a_1a_2}(M))^{\mathrm{T}}$$

be the population counterpart of $\boldsymbol{\gamma}$. Let

$$\eta_2(u, v) = \partial\eta(u, v)/\partial v .$$

Suppose that $\rho_{a_1a_2}(l) = \rho$ if $l = 0$ and zero if $l \neq 0$. This would be a realistic assumption with many economic time series and is related to the so-called seemingly unrelated regression problem. Let

$$\sqrt{n}\hat{\boldsymbol{\gamma}}_1 = \sqrt{n}(\hat{\gamma}_{a_1a_2}(-1), \ldots, \hat{\gamma}_{a_1a_2}(-M))^{\mathrm{T}}$$

and

$$\sqrt{n}\hat{\boldsymbol{\gamma}}_2 = \sqrt{n}(\hat{\gamma}_{a_1a_2}(1), \ldots, \hat{\gamma}_{a_1a_2}(M))^{\mathrm{T}} .$$

Using the theorem of Li and Hui (1994) and after some algebra the asymptotic covariance matrices $\mathbf{P}_h = (h = 1, 2)$ for $\sqrt{n}\hat{\boldsymbol{\gamma}}_h$ can be shown to be

$$\mathbf{P}_h = \boldsymbol{G}_h + (\tau_h^2 - 2K_h\tau_h)\tilde{\boldsymbol{X}}_h\boldsymbol{V}_h\tilde{\boldsymbol{X}}_h^{\mathrm{T}} , \tag{4.8}$$

where $\boldsymbol{G}_h = a\mathbf{1}_M$ with $a = E[\eta(a_{1t}/\sigma_1, a_{2t'}/\sigma_2)^2]$ and $\mathbf{1}_M$ the $M \times M$ identity matrix, $t \neq t'$; $\tau_h = \sigma_h^{-1}E[\eta_h(a_{1t}/\sigma_1, a_{2t'}/\sigma_2)a_{ht'}]$, $t \neq t'$; and $\tilde{X}_h = (x_{ijh}) = (\phi'_{h,i-j})^{\mathrm{T}}$, $K_h = \beta_h E[\eta(a_{1t}/\sigma_1, a_{1t'}/\sigma_1)\eta(a_{1t}/\sigma_1, a_{2t'}/\sigma_2)]$ $/(\alpha_h\sigma_h)$, $t \neq t'$. In general a, τ_h and K_h are unknown but can be estimated consistently by sample averages. To test the null hypotheses

$$H_0^{(1)} : \rho_{a_1a_2}(-i) = 0 \ ; \quad H_0^{(2)} : \rho_{a_1a_2}(i) = 0 \ , \quad i = 1, \ldots, M \ ,$$

against the simple negation of $H_0^{(1)}$ or $H_0^{(2)}$ when $\rho \neq 0$, the following statistics analogous to McLeod (1979) are suggested:

$$Q_h^*(M) = n\hat{\boldsymbol{\gamma}}_h^{\mathrm{T}}\hat{\mathbf{P}}_h^{-1}\hat{\boldsymbol{\gamma}}_h \ , \quad h = 1, 2, \tag{4.9}$$

where $\hat{\mathbf{P}}_h$ denotes the matrix $\hat{\mathbf{P}}_h$ evaluated using the residual autocovariance estimates $\hat{\phi}_h$. Under $H_0^{(h)}$: $Q_h^*(M)$ is asymptotically chi-squared with M degrees of freedom. Thus the result in Chapter 3 is robustified. If $\rho_{a_1a_2}(0) = 0$, then $\tau_h = 0$ and robustified versions of Haugh's $P(S)$ tests are obtained.

Li and Hui (1994) considered some small simulation experiments to study the effect of outliers on the size and power of the $Q_h^*(M)$ statistics. The corresponding unrobustified statistics from McLeod (1979) $Q_h(M)$ were also included in the study. The statsitic $Q_h(M)$ is just (4.9) evaluated using the conditional least squares estimates. The two time series processes

Table 4.2 *Empirical means, variances, and upper significance levels of* Q_i^* *and* Q_i, $i = 1, 2$. $M = 10$. $n = 100$. *Bracketed values correspond to the no outlier situation (Li and Hui, 1994). Reproduced with the permission of Taylor & Francis Ltd.*

	Mean	Variance	Upper 10%	Upper 5%
(Bisquares)				
Q_1^*	9.58	19.07	0.080	0.048
	(9.53)	(18.62)	(0.090)	(0.038)
Q_2^*	9.54	20.93	0.082	0.048
	(9.57)	(20.86)	(0.086)	(0.040)
(Huber's)				
Q_1^*	9.05	16.23	0.052	0.020
	(9.64)	(18.65)	(0.074)	(0.038)
Q_2^*	9.16	14.40	0.050	0.018
	(9.81)	(18.29)	(0.010)	(0.050)
(Least squares)				
Q_1	5.48	6.30	0.004	0.002
	(9.78)	(16.64)	(0.080)	(0.032)
Q_2	4.05	2.72	0.000	0.000
	(10.14)	(16.17)	(0.088)	(0.032)

were assumed to be autoregressive of order one and the $\{a_{h,t}\}$ processes were instantaneously correlated with correlation ρ. They were generated as Gaussian variates. The autoregressive parameters $\phi_{h1}(h = 1, 2)$ have a value of 0.5. The value of ρ was 0.3, the variances $\sigma_h^2 = 1(h = 1, 2)$, and $M = 10$. Yule-Walker estimates were used for $Q_h(M)$. Mallows type η were used with bisquares and Huber ψ functions. The tuning constant of the bisquares function was 5.58 and that of Huber's function was 1.65 (Bustos and Yohai, 1986). In Table 4.2 the empirical mean, variance, and the number of rejections at the upper 5 and 10% significance levels of a chi-squared distribution with ten degrees of freedom were reported in the two situations corresponding to the respective presence and absence of outliers. The outliers situations were created by adding a value of ten to the 26th and 51st positions of the first series and the same value to the 51st and 76th positions of the second. There were 500 replications each of length $n = 100$ for each case. In Table 4.2 it can be seen that where there were no outliers, the finite sample distributions of $Q_h^*(M)$ and $Q_h(M)$ matched the asymptotic chi-squared distribution fairly well. However, with outliers the unrobustified statistics became rather conservative. Note that the $Q_h^*(M)$ statistics were more robust than the $Q_h(M)$ statistics in all aspects. However, the $Q_h^*(M)$ statistics based on the bisquares gave the best set of results. All the corresponding entries in Table 4.2 gave values very close to that of a chi-squared distribution with ten degrees of freedom. Li and Hui (1994) studied also the power of the tests. The data generating processes were

$$X_{1t} = 0.5X_{1t-1} + \theta_{11}a_{2t-1} + a_{1t} \ , \ X_{2t} = 0.5X_{2t-1} + \theta_{21}a_{2t-1} + a_{2t} \ .$$

The values of of $(\theta_{11}, \theta_{21})$ were $(0.15, 0.15)$, $(0.30, 0.30)$, and $(0.50, 0.50)$.

Table 4.3 *Empirical power for Q_i^* and Q_i, $i = 1, 2$. $M = 10$. Entries are number of rejections in 500 replications at the nominal upper 5 and 10% critical values of a chi-square distribution of 10 degrees of freedom. Bracketed values correspond to the no outlier situation (Li and Hui, 1994). Reproduced with the permission of Taylor & Francis Ltd.*

$\theta_{11} = \theta_{12}$	0.15		0.30		0.50	
(Bisquares)	Upper 10%	5%	10%	5%	10%	5%
Q_1^*	75	39	173	118	323	250
	(78)	(41)	(210)	(135)	(385)	(309)
Q_2^*	79	42	171	122	327	250
	(96)	(55)	(108)	(146)	(367)	(305)
(Huber's)						
Q_1^*	41	20	90	52	208	138
	(82)	(46)	(204)	(135)	(374)	(300)
Q_2^*	39	21	94	59	226	153
	(89)	(55)	(202)	(133)	(358)	(309)
(Least squares)						
Q_1^*	8	4	22	16	91	59
	(96)	(46)	(264)	(191)	(434)	(398)
Q_2^*	4	3	15	8	84	54
	(90)	(49)	(255)	(181)	(437)	(399)

For simplicity $E(a_{1t}a_{2t}) = 0$. There were again 500 replications each with length 100. The two time series were modeled independently as univariate AR(1) processes. The $Q_h^*(10)$ and $Q_h(10)$ statistics were applied to the residuals. The outlier situation was created in the same way as in the first experiment. The results of the power study are recorded in Table 4.3. With no outliers present the performances of the Q_h^* statistics in general were respectable but somewhat less powerful than those of Q_h's. As in the first experiment the Q_h statistics fell off rapidly where there were just two outliers in each of the series. Their power was almost zero unless the θ_{h1} were very large. Again the Q_h^* statistics based on the Huber type psi function performed much better but the overall best performers were the Q_h^* statistics based on the bisquares. Comparatively very little fall off in performance was observed across the parameter range considered. The Q_h^* statistics based on the bisquares are recommended for actual use in place of the Q_h statistics if outliers are suspected to be present. The robustified residual cross-correlations and the statistics $Q_h^*(M)$ can be easily computed from the RA estimates. Duchesne and Roy (2003) extended further Li and Hui's result by robustifying a class of tests proposed by Hong (1996a).

4.3 A robust estimation method for vector time series

Let $\mathbf{X}_t = (x_{1t}, \ldots, x_{lt})^{\mathrm{T}}$ be an l-dimensional stationary time series observed over time period $t = 1, \ldots, n$. Li and Hui (1989) proposed an estimator of the autoregressive parameters that is sturdy against contamination of the AO type where the observations x_{it} $(i = 1, \ldots, l; t = 1, \ldots, n)$ are replaced by $x_{it} + \delta_{it}$, where the quantities $\{\delta_{it}\}$ are unobservable. Suppose that the process $\mathbf{X}_t$ satisfies the pth-order autoregression

$$(\mathbf{1}_l - \boldsymbol{\phi}_1 B - \cdots - \boldsymbol{\phi}_p B^p)(\mathbf{X}_t - \boldsymbol{\mu}) = \boldsymbol{a}_t \ , \tag{4.10}$$

where B denotes the backward shift operator; $\mathbf{1}_l$ is the $l \times l$ identify matrix; $\boldsymbol{\phi}_i$ are $l \times l$ autoregressive parameters; $\boldsymbol{\mu}$ is a $l \times 1$ vector of constants; and a_t are independent l-dimensional white noise with mean zero and covariance matrix $\boldsymbol{\Delta}$. For stationarity it is required that all roots of $\det(\mathbf{1}_l - \boldsymbol{\phi}_1 B - \cdots - \boldsymbol{\phi}_p B^p)$ lie outside the unit circle.

Denote by $A \otimes B$ the Kronecker product of the matrices A and B. Let $\mathrm{vec}(\cdot)$ be the column vectorizing operation. Suppose for simplicity $\boldsymbol{\mu} = \mathbf{0}$. Let $\boldsymbol{\phi} = (\boldsymbol{\phi}_1, \ldots, \boldsymbol{\phi}_p)$ and $Z_{t-1}^{\mathrm{T}} = (X_{t-1}^{\mathrm{T}}, \ldots, X_{t-p}^{\mathrm{T}})$. Suppose that a_t is Gaussian; then the conditional estimator of $\boldsymbol{\beta} = \mathrm{vec}(\boldsymbol{\phi}^{\mathrm{T}})$ is obtained by minimizing the quantity $S = \frac{1}{2} \sum \boldsymbol{a}_t^{\mathrm{T}} \boldsymbol{\Delta}^{-1} \boldsymbol{a}_t$, where the sum is over $t = p+1, \ldots, n$ (Wilson, 1973). Since (4.10) can be rewritten as $\mathbf{X}_t - \mathrm{vec}(Z_{t-1}^{\mathrm{T}} \boldsymbol{\phi}^{\mathrm{T}}) = \boldsymbol{a}_t$ and $\mathrm{vec}(Z_{t-1}^{\mathrm{T}} \boldsymbol{\phi}^{\mathrm{T}}) = (\mathbf{1}_l \otimes Z_{t-1}^{\mathrm{T}}) \boldsymbol{\beta}$,

$$\partial S / \partial \boldsymbol{\beta} = \sum (\mathbf{1}_l \otimes Z_{t-1}) \boldsymbol{\Delta}^{-1} \boldsymbol{a}_t \ .$$

Using the result $\mathrm{vec}(ABC) = (C^{\mathrm{T}} \otimes A) \mathrm{vec}\, B$ repetitively the above can be written as

$$\partial S / \partial \boldsymbol{\beta} = (\boldsymbol{\Delta}^{-1} \otimes \mathbf{1}_{lp}) \sum \boldsymbol{a}_t \otimes Z_{t-1} .$$

Let $(\mathbf{1}_l - \boldsymbol{\phi}_1 B - \cdots - \boldsymbol{\phi}_p B)^{-1} = \sum_i \boldsymbol{\psi}_i B^i$. It can be seen that

$$(\boldsymbol{a}_t \otimes Z_{t-1})^{\mathrm{T}} = (a_{1t}\mathbf{X}_{t-1}^{\mathrm{T}}, \ldots, a_{lt}\mathbf{X}_{t-p}^{\mathrm{T}})$$
$$= \left(\sum a_{1t}\boldsymbol{a}_{t-1-i}^{\mathrm{T}}\boldsymbol{\psi}_i^{\mathrm{T}}, \ldots, \sum a_{lt}\boldsymbol{a}_{t-p-i}^{\mathrm{T}}\boldsymbol{\psi}_i^{\mathrm{T}}\right) .$$

Since Δ^{-1} is nonsingular and can be estimated separately using residuals from the estimation of $\boldsymbol{\beta}$ as is the case with the scale parameter in the univariate case, the estimating equation for $\boldsymbol{\beta}$ can be written as

$$\sum_t \mathbf{1}_l \otimes \left(\mathbf{1}_p \otimes \sum_{i=0}^{\infty} \boldsymbol{\psi}_i B^i\right)\{\boldsymbol{a}_t \otimes (\boldsymbol{a}_{t-1}^{\mathrm{T}}, \ldots, \boldsymbol{a}_{t-p}^{\mathrm{T}})^{\mathrm{T}}\} = 0 . \qquad (4.11)$$

Alternatively (4.11) can be written more simply as in Li and Hui (1989),

$$\sum_t \sum_i \boldsymbol{\psi}_i a_{h,t}\boldsymbol{a}_{t-j-i} = 0 \quad (j = 1, \ldots, p; h = 1, \ldots, l) ,$$

where $a_{h,t} = 0$ for $t < p + 1$.

Motivated by the univariate result we robustify the products $a_{h,t}a_{k,t'}$, by a bounded and continuous function $\eta(u, v)$ that is odd in each variable. As before the two possible choices for $\eta(\cdot, \cdot)$ are $\eta(u, v) = \psi(u)\psi(v)$ or $\eta(u, v) = \psi(uv)$, where ψ is a bounded and continuous odd function. The former choice is said to be of Mallows type and the latter of Hampel type. The function ψ can be in the Huber family or the bisquares family. Let

$$\eta(a_{h,t}, a_{t-j}) = [\eta(a_{h,t}, a_{1,t-j}), \ldots, \eta(a_{h,t}, a_{l,t-1})]^{\mathrm{T}} ,$$
$$\delta_{h,j,t} = \sum_i \boldsymbol{\psi}_i \eta(a_{h,t}, a_{t-j-i}) , \quad \delta_{h,t} = (\delta_{h,1,t}, \ldots, \delta_{h,p,t})^{\mathrm{T}} .$$

The estimating equations can then be written as

$$L = \sum_t \delta_t = 0 , \qquad (4.12)$$

where

$$\delta_t^{\mathrm{T}} = [(\delta_{1,1,t}, \ldots, \delta_{1,p,t}), \ldots, (\delta_{l,1,t}, \ldots, \delta_{l,p,t})] .$$

Now, define (Bustos and Yohai, 1986 and Li, 1988)

$$\gamma_{h,k}(j) = \sum_{t=p+1+j}^{n} \eta(a_{h,t}a_{k,t-j}) ,$$

and $\boldsymbol{\gamma}_h(j) = (\gamma_{h,1}(j), \ldots, \gamma_{h,l}(j))^{\mathrm{T}}$; then (4.12) can be written (Li and Hui, 1989)

$$\sum_{i=0}^{n-j-p-1} \boldsymbol{\psi}_i \boldsymbol{\gamma}_h(i + j) = 0 \quad (j = 1, \ldots, p; h = 1, \ldots, l) . \qquad (4.13)$$

Clearly (4.13) reduces to the univariate residual autocovariance estimating equations when $l = 1$. A routine for nonlinear equations can then be used to obtain $\hat{\boldsymbol{\beta}}$. Such estimators will be called the multivariate residual autocovariance estimators.

If $\boldsymbol{\mu}$ is not zero, then the series $\mathbf{X}_t$ may first be robustly centered, say, by using α-trimmed means or similar robust location estimators. Alternatively $\boldsymbol{\beta}, \boldsymbol{\mu}$, and $\boldsymbol{\Delta}$ may be estimated jointly (Bustos, Fraiman, and Yohai, 1984) by applying the results of Maronna (1976). If $\eta(u, v)$ is of the Mallows type then, as in the univariate case, an iterative least squares scheme for $\hat{\boldsymbol{\beta}}$ can be used which will in general save computer time.

Let

$$A_{hh'} = E\{\eta(a_{h,t}, a_{t'})\eta^{\mathrm{T}}(a_{h't}, a_{t'})\} \ . \tag{4.14}$$

Let the robustified residual autocovariances at lag j be a $l \times l$ matrix C_j with (g, h)th element $\sum \eta(a_{g,t}, a_{h,t-j})/n$. It can be seen that

$$n \ \mathrm{cov}\{\mathrm{vec}(C_j^{\mathrm{T}})\} = (A_{km}) \ , \quad (k, m = 1, \ldots, l) \ . \tag{4.15}$$

Let $\boldsymbol{C} = \mathrm{vec}\{(C_1, \ldots, C_M)^{\mathrm{T}}\}$, where $0 \ll M \ll n$, then $\sqrt{n}\boldsymbol{C}$ can be shown to be asymptotically distributed with mean zero and covariance matrix $\boldsymbol{\Omega}$, where $\boldsymbol{\Omega} = (P_1^{\mathrm{T}}, \ldots, P_l^{\mathrm{T}})$ with $P_i = (\mathbf{1}_M \otimes A_{i1}, \ldots, \mathbf{1}_M \otimes A_{il})$ $(i = 1, \ldots, l)$. The quantity $Q'_M = n\hat{\boldsymbol{C}}^{\mathrm{T}}\boldsymbol{\Omega}^{-1}\hat{\boldsymbol{C}}$ can be shown to be asymptotically chi-squared with degrees of freedom $(M - p)l^2$. In practice $\boldsymbol{\Omega}$ can be replaced by a consistent estimate $\hat{\boldsymbol{\Omega}}$. As in the Gaussian situation some adjustment to Q'_M is desirable. One possible adjustment is by adding the quantity $\frac{1}{2}l^2M(M+1)/n$ to Q'_M (see (3.10)). For simplicity we use Q'_M to denote also the adjusted statistic below.

Example 4.1 *The mink-mustrat data. (Li and Hui, 1989). ©1989 Biometrika Trust, reproduced with the permission of Oxford University Press*

The proposed estimation procedure and the robustified goodness-of-fit statistic was applied to the mink-muskrat data (1848–1911) which have been studied by Chan and Wallis (1978), Nicholls (1979), Tong (1983), and Heathcote and Welsh (1988) using the functional least squares approach. Several of these authors have considered a first order autoregressive model but Tong (1983) gave evidence that the series may be nonlinear. Denote by x_{1t} the first differences of the logarithm of the muskrat data and x_{2t} the logarithm of the mink series. Let $\mathbf{X}_t = (x_{1t}, x'_{2t})^{\mathrm{T}}$. It is believed that observations 39 and 61 in the first series and observations 4, 38, and 42 in the second may be outliers. A first order autoregression was fitted to $\mathbf{X}_t$ using the residual autocovariances estimation procedure. A Mallows type η function with a Huber ψ function were used. The constant c in the Huber ψ function was chosen to be $c'\hat{\sigma}_i$ where $\hat{\sigma}_i$ was a *robust scale estimate of the argument* $u = \hat{a}_{it}$ $(i = 1, 2)$. *Since there were not too many suspected outliers a choice of* $c' = 2{\cdot}0$ *was used allowing a moderate amount of protection. The scale parameters* $\hat{\sigma}_i$ *were computed using the median of* $(|\hat{a}_2|, \ldots, |\hat{a}_n|)/0{\cdot}6745$ *during each iteration. A routine for systems of nonlinear equations such as the* IMSL *subroutine* ZSCNT *can be used but since we have a Mallows type* η *function, the iterative scheme suggested at the end of §4.1 was used. The* IMSL *subroutine* LLSQF *was used to obtain the estimates. The mink series was centered by a 40% trimmed mean (Heathcote and Welsh, 1988). The robustified portmanteau statistics* Q'_M, $M = 20$, *was also computed. The least squares estimates and the unrobustified portmanteau statistics in Chapter 3* $Q^*(M)$ *were also computed for comparison. Here* x_{2t} *is centered around the sample mean. The results are as follows. For least squares estimates*

$$vec(\hat{\boldsymbol{\phi}}_1) = (0{\cdot}036, 0{\cdot}310, -0{\cdot}581, 0{\cdot}786)^{\mathrm{T}}, \ vech(\hat{\boldsymbol{\Delta}}) = (0{\cdot}083, 0{\cdot}016, 0{\cdot}072)^{\mathrm{T}}$$

and $Q^*(20) = 128{\cdot}5$. *For residual autocovariances estimates*

$$vec(\hat{\boldsymbol{\phi}}_1) = (0{\cdot}022, 0{\cdot}310, -0{\cdot}574, 0{\cdot}789)^{\mathrm{T}}, \ vech(\hat{\boldsymbol{\Delta}}) = (0{\cdot}073, 0{\cdot}012, 0{\cdot}058)^{\mathrm{T}}$$

and $Q'_{20} = 124{\cdot}3$. *The residual autocovariances estimate of* ϕ_1 *is closer to the ordinary least squares estimates than the functional least squares estimates (Heathcote and Welsh, 1988). The effect of outliers also seems to be small. However, Heathcote and Welsh considered the data from 1848 to 1909 only. The two portmanteau statistics are also very close. They suggested that under the assumption of linearity the first order autoregressive model is probably inadequate contrary to the claim of Chan and Wallis (1978).*

4.4 The trimmed portmanteau statistic

A common technique in robust statistical estimation is by means of trimming. See Lo and Li (1990) and the references therein. Chan (1994) proposed a robust portmanteau test based on trimming. It seems that trimming is also useful in strengthening the resistance of a statistic to extreme value. The $\hat{r}_k$ in Q_m (2.11) is replaced by the α-trimmed residual autocorrelation which is an extension of the trimmed sample autocorrelation proposed by Chan and Wei (1992). Let $\hat{a}_{(p+1)} \leq \hat{a}_{(p+2)} \leq \cdots \leq \hat{a}_{(n)}$ be the ordered residuals from an estimated ARMA model. The α-trimmed residual autocorrelation function is defined by

$$\hat{\rho}_k^{(a)} = \frac{\hat{\gamma}_k^{(\alpha)}}{\hat{\gamma}_0^{(\alpha)}}, \tag{4.16}$$

where

$$\hat{\gamma}_k^{(a)} = \frac{1}{\sum_{t=p+k+1}^{n} L_{t-k}^{(\alpha)} L_t^{(\alpha)}} \left(\sum_{t=p+k+1}^{n} \hat{a}_{t-k}\hat{a}_t L_{t-k}^{(\alpha)} L_t^{(\alpha)} \right) ,$$

and

$$L_t^{(\alpha)} = \begin{cases} 0 & \text{if } \hat{a}_t \leq \hat{a}_{(g)} \text{ or } \hat{a}_t \geq \hat{a}_{(n-g+1)} , \\ 1 & \text{otherwise} , \end{cases}$$

for $p+1 \leq t \leq n$ with g is the integer part of $[\alpha n]$ and $0 \leq \alpha < 0.5$.

Define

$$C_L(k) = \frac{1}{n} \sum_{t=p+k+1}^{n} L_{t-k}^{(\alpha)} L_t^{(\alpha)} ,$$

and assume that the limits

$$\lim_{n\to\infty} C_L(k) = \nu_k \quad \text{a.s.}$$

exist for all finite k. Let

$$Q_m^{(\alpha)} = \sum_{k=1}^{m} (n\nu_k)[\hat{\rho}_k^{(\alpha)}]^2 . \tag{4.17}$$

The quantity ν_k is not known in general but it can be replaced by

$$\hat{\nu}_k = \frac{1}{n} \sum_{t=p+k+1}^{n} L_{t-k}^{(\alpha)} L_t^{(\alpha)} .$$

Let

$$\widehat{\mathbf{\Upsilon}}_m^{(\alpha)} = \left(\hat{\rho}_1^{(\alpha)}, \ldots, \hat{\rho}_m^{(\alpha)}\right)^{\mathrm{T}} .$$

Following Marshall (1980) and Dunsmuir and Robinson (1981), Chan (1994) showed that the asymptotic distribution of $\sqrt{n}\widehat{\mathbf{\Upsilon}}_m^{(\alpha)}$ is Gaussian with mean zero and covariance matrix

$$\nu_k^{-1}(\mathbf{1}_m - \mathbf{X}\mathbf{I}^{-1}\mathbf{X}^{\mathrm{T}}) . \tag{4.18}$$

It follows at once from the classical result (McLeod, 1978) that, if $n \gg m$ and the model is adequate, then $\sqrt{n\nu_k}\widehat{\mathbf{\Upsilon}}_m^{(\alpha)}$ has an asymptotic covariance matrix that is idempotent of rank $(m-p-q)$. Hence, the α-trimmed portmanteau statistic in (4.17) is asymptotically distributed as chi-squared with degrees of freedom $m-p-q$. A simulation study by Chan (1994) showed that the adjustment factor $n/(n-k)$ is not necessary in this situation. It might be due to the fact that the ν_k have already provided some adjustment for the lag effects. Chan also showed by a small simulation study that $Q_m^{(\alpha)}$ is more powerful than $\tilde{Q}_m$ of §4.1 under additive outliers while $\tilde{Q}_m$ is more powerful than $Q_m^{(\alpha)}$ under innovative outliers.

CHAPTER 5

Nonlinear models

5.1 Introduction

Toward the end of the seventies of the last century there was an increasing demand to model more complex time series features than those given by a linear autoregressive moving average (ARMA) structure. One drawback of the stationary ARMA model with Gaussian noise a_t is that it is unable to capture time irreversibility. Time irreversibility is one of the major features exhibited by a nonlinear or non-Gaussian time series model. A stationary time series X_t is time reversible if for any integer $n > 0$, and any $t_1, t_2, \ldots, t_n$ that are integers, the vectors $(X_{t_1}, X_{t_2}, \ldots, X_{t_n})$ and $(X_{-t_1}, X_{-t_2}, \ldots, X_{-t_n})$ have the same multivariate distribution. A stationary time series that is not time reversible is said to be time irreversible. The result of Weiss (1975) showed that stationary ARMA processes with a nontrivial AR component are time reversible if and only if they are Gaussian. The technical report by Tong and Zhang (2003) gave more results on the conditions of time reversibility. Figure 5.1 shows the time series plot of the Canadian Lynx data 1821–1934 as listed by Elton and Nicholson (1942). It can be seen that the time series take more time to reach the peaks than to come down from the peaks to the troughs. This suggested that the above definition of reversibility would not hold for the Lynx data. Another way of seeing this is to place a mirror on the y-axis and for the mirror image it will take less time to climb up to the peaks than to come down from the peaks. Naturally, new nonlinear models are required to capture these kinds of features. There are, of course, other features arising from nonlinearity that cannot be mimicked by the linear Gaussian ARMA models. One of these is the limit cycles exhibited by a nonlinear difference equation. A limit cycle is a set of points $\{x_1, \ldots, x_T\}$ with a mapping $f(x)$ such that $f(x_i) = x_{i+1}$, $i = 1, \ldots, T-1$, and $x_{T+i} = x_i$, $i = 1, 2, \ldots$. Suppose a time series is defined by $X_t = g(X_{t-1}, a_t)$, where a_t is a zero mean white noise process independent of X_{t-1}. Then we say that X_t admits a limit cycle, when a_t is set to its mean zero, the mapping $X_t = g(X_{t-1}, 0)$ induces a recursion $X_t = f(X_{t-1})$ that has a limit cycle as $t \to \infty$ (Chan and Tong, 1990). A stationary ARMA model can only have a limit cycle in the trivial case $T = 1$.

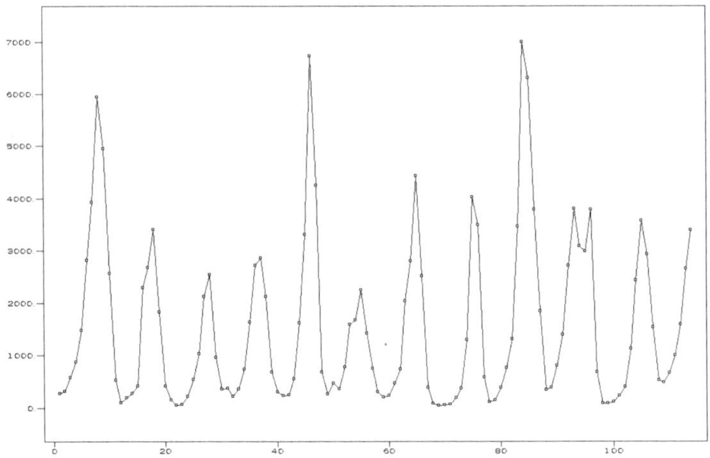

Figure 5.1 *Sample path of the Canadian Lynx data*

Two major classes of nonlinear models were developed by the end of the 1970s. These were the threshold model of Tong (1978) and the bilinear models of Granger and Andersen (1978). A full generalization of the threshold model occurred in Tong and Lim (1980) and a full generalization of the bilinear model appeared in Subba Rao (1981). In its simplest form the threshold autoregressive model of order 1 is defined by

$$\begin{aligned} X_t &= \phi X_{t-1} + a_t \ , \quad \text{if} \quad X_{t-1} > C \\ X_t &= \phi' X_{t-1} + a'_t \ , \quad \text{if} \quad X_{t-1} \leq C \ , \end{aligned} \tag{5.1}$$

where C is the threshold value, $\phi \neq \phi'$, a_t is white noise with mean 0 and variance σ_0^2, while a'_t is white noise with mean 0 and variance σ_1^2. Intuitively, the time series X_t satisfies a different autoregression or regime whenever the threshold C is crossed. Many hydrological series appear to satisfy this model. For example if there is a large amount of precipitation then it seems reasonable to assume that a river flow series will behave quite differently.

For stationarity of the model (5.1) it is required that $\phi_1 < 1$, $\phi_2 < 1$, $\phi_1\phi_2 < 1$ (Chan, Petruccelli, Tong, and Woolford, 1985). Equation (5.1) can be easily fitted by the least squares method if C is known. Various proposals have been made on the estimation of C when it is unknown. One approach is to use as candidates of C a subset of the order statistics of the realization $X_1, \ldots, X_n$. An information criterion such as the

Akaike information criterion (AIC) or the Bayesian information criterion (BIC) can then be used to pick an estimate for C from the subset. Chan (1993) showed that the estimate of C is in fact super-consistent in the sense that its estimate $\hat{C}$ converges at a rate of $1/n$ to the true value. This is faster than the usual rate of $n^{-1/2}$. The model (5.1) can be generalized in many ways. For example, more than one threshold value can be considered so that there will be more than two regimes. For ease of exposition we will work with only two regimes in this book. A general 2-regime threshold autoregressive TAR model can be defined as

$$\begin{aligned} X_t &= \phi_0 + \phi_1 X_{t-1} + \cdots + \phi_{p_1} X_{t-p_1} + a_t \ , \quad \text{if} \quad X_{t-d} > C \\ X_t &= \phi_0' + \phi_1' X_{t-1} + \cdots + \phi_{p_2}' X_{t-p_2} + a_t' \ , \quad \text{otherwise} \ , \end{aligned} \tag{5.2}$$

where a_t is $(0, \sigma_0^2)$ white noise and a_t' is white noise with mean 0 and variance σ_1^2, $1 \le d \le \max(p_1, p_2)$. Tong and Lim (1980) called (5.2) the self-exciting threshold autoregression (SETAR) model. Clearly, without loss of generality we can assume $p_1 = p_2 = p$ by setting some of the $\phi' s$ to 0. Again, least squares estimation can be done easily given d and C. Let $D = \{1, \ldots, p\}$ and $\mathcal{C} = \{X_{(1)}, \ldots, X_{(n)}\}$ where $X_{(i)}$ are the order statistics of X_i, $i = 1, \ldots, n$. The estimation of d and C can be based on an information criterion such as AIC or BIC applied on elements of D and a subset of $\mathcal{C}$. (To make sure that there will be enough observations in each regime we will have to use only observations between say, the 20th and the 80th percentile.) TAR models can easily model features like limit cycles and time irreversibility (Tong and Lim, 1980). Because of these and its piecewise linear nature the TAR model is now a rather successful nonlinear model.

Another important class of models, the bilinear models, were considered by Subba Rao (1977) and by Granger and Andersen in their 1978 monograph. In the simplest case a bilinear model takes the form

$$X_t = \beta X_{t-l} a_{t-k} + a_t \ , \tag{5.3}$$

where β is a parameter, $k \ge 1$, $l \ge 1$, and a_t is white noise with mean 0 and variance σ^2. Model (5.3) is both strictly and covarivance stationary if $\beta^2 \sigma^2 < 1$ (Pham and Tran, 1981). Properties of X_t depend on whether $k > l$, $k = l$, or $k < l$. When $l > k$ it is called the superdiagonal model, $k = l$ the diagonal model, and $k > l$ the subdiagonal model (Granger and Andersen, 1978). When $l > k$ the autocorrelations for X_t are all zero and hence X_t would be mistaken as white noise if only autocorrelations were inspected for a dependence structure. This can be seen as follows. First observe that since $k < l$,

$$\begin{aligned} E(X_t) &= \beta E(X_{t-l} a_{t-k}) + E(a_t) \\ &= 0 \ . \end{aligned}$$

Hence, the lag i autocovariance, $i \geq 1$, is

$$\begin{aligned} E(X_t X_{t-i}) &= \beta^2 E(X_{t-l} a_{t-k} X_{t-l-i} a_{t-k-i}) + \beta E(X_{t-l} a_{t-k} a_{t-i}) + \\ &\quad + \beta E(X_{t-l-i} a_{t-i-k} a_t) + E(a_t a_{t-i}) \\ &= 0 \,. \end{aligned}$$

The above is true because inside each bracket at least one of the a_t's has a time index larger than all the other variables. Similarly, assuming stationarity up to the fourth order we can show that X_t^2 are correlated. See Li (1984) for more results of this kind.

A general bilinear model of order (p, q, P, Q) can be defined as

$$X_t = \sum_{j=1}^{p} \phi_j X_{t-j} + \sum_{j=1}^{q} \theta_j a_{t-j} + \sum_{k=0}^{Q} \sum_{l=1}^{P} \beta_{kl} a_{t-k} X_{t-l} + a_t \,. \qquad (5.4)$$

It is obvious from (5.4) that there is a large number of parameters for this general bilinear model. Subba Rao (1981) gave more details on (5.4) and its estimation which would have to be based on the Newton-Raphson method. Estimation and model selection could be problematic. Stationarity conditions have been considered by many authors, for example, Liu (1992) and Liu and Brockwell (1988). Terdik (1999) gave an updated discussion on bilinear models via the frequency domain approach.

A general class of nonlinear model that can be considered as encompasssing both the threshold and bilinear models is the state dependent model of Priestley (1980, 1988). Let the model for X_t be given by

$$X_t = g(X_{t-1}, \ldots, X_{t-p}, a_{t-1}, \ldots, a_{t-q}) + a_t \,.$$

Suppose that g is known and is analystic, then using a first order Taylor expansion about $(X_{t_0-1}, \ldots, X_{t_0-p}, a_{t_0-1}, \ldots, a_{t_0-q})^{\mathrm{T}} = \boldsymbol{x}_{t_0-1}$, we have

$$\begin{aligned} X_t &= g(X_{t_0-1}, \ldots, a_{t_0-q}) + \sum_{i=1}^{p} g_i(\boldsymbol{x}_{t-1})(X_{t-u} - X_{t_0-u}) \\ &\quad + \sum_{i=1}^{q} h_j(\boldsymbol{x}_{t-1})(a_{t-u} - a_{t_0-u}) + a_t \end{aligned} \qquad (5.5)$$

where $\boldsymbol{x}_t$ is called the state vector and $g_i = \partial h/\partial X_{t-i}$ and $h_i = \partial h/\partial a_{t-i}$. We note that (5.5) can be rewritten in the following general form,

$$X_t - \sum_{i=1}^{p} \phi_i(\boldsymbol{x}_{t-1}) X_{t-u} = \mu(\boldsymbol{x}_{t_0-1}) + a_t + \sum_{i=1}^{q} \theta_i(\boldsymbol{x}_{t-1}) a_{t-i} \,. \qquad (5.6)$$

We call (5.6) a state-dependent model (SDM) of order (p, q). It can be seen that the ARMA(p, q) model is a special case of (5.6) by requiring $\phi_i(\boldsymbol{x}_{t-1})$, $\theta_i(\boldsymbol{x}_{t-1})$, and $\mu(\boldsymbol{x}_{t-1})$ to be constants. We have a bilinear model if $\mu(\boldsymbol{x}_{t-1})$ and $\phi_i(\boldsymbol{x}_{t-1})$ are constants but $\theta_i(\boldsymbol{x}_{t-1}) = b_{ij}X_{t-j}$, say. A threshold model results if in (5.6) all $\theta_i = 0$; $\mu(\boldsymbol{x}_{t_0-1}) = \phi_0$, $\phi_i(\boldsymbol{x}_{t-1}) = \phi_i$, if $X_{t-d} > C$ and $\mu(\boldsymbol{x}_{t_0-1}) = \phi_0'$, $\phi_i(\boldsymbol{x}_{t-1}) = \phi_i'$, if $X_{t-d} \leq C$. If $\mu(\boldsymbol{x}_{t_0-1}) = \theta_i(\boldsymbol{x}_{t-1}) = 0$, $\phi_i(\boldsymbol{x}_{t-1}) = \phi_i + \pi_i e^{-\gamma x_{t-1}^2}$, we have the exponential autoregressive model of Ozaki (1980) and Haggan and Ozaki (1981). A state-space representation for (5.6) can be constructed as in Priestley (1988). The state-space representation facilitiates model identification and estimation. Note that by allowing $\phi_i(\boldsymbol{x}_{t-1})$ and $\theta_i(\boldsymbol{x}_{t-1})$ to be arbitrary functions of t results in a non-stationary model. For example, let

$$\begin{aligned}\phi_i(\boldsymbol{x}_{t-1}) &= \phi_i^0 + \boldsymbol{x}_t^{\mathrm{T}}\boldsymbol{\gamma}_i^{\mathrm{T}} \,, \\ \theta_i(\boldsymbol{x}_{t-1}) &= \theta_i^0 + \boldsymbol{x}_t^{\mathrm{T}}\boldsymbol{\beta}_i^{\mathrm{T}}\end{aligned}$$

and allow $\boldsymbol{\gamma}_i$ and $\boldsymbol{\beta}_i$ to wander like random walks. Readers are referred to Priestley (1988) for a thorough discussion on SDM models.

5.2 Tests for general nonlinear structure

It seems natural that given a time series realization one should first consider fitting a linear ARMA model to the data before entertaining a nonlinear model. This is both from a practical point of view and from the fact that any confounding effect with linearity should be avoided. Indeed many tests for nonlinearity are valid only after this linear modeling step has been taken.

In what follows X_t is assumed to be stationary up to the fourth order. A test for nonlinearity with infinite variance has been considered by Resnick and van Den Berg (2001). Their treatment is beyond the scope of this monograph.

(i) McLeod-Li test

We first consider the general portmanteau type test for nonlinearity by McLeod and Li (1983). Let X_t be a fourth-order stationary time series: Given a realization of X_t, $t = 1, \ldots, n$, an appropriate ARMA model (2.1) is first fitted to the data. Let $\hat{a}_t$ be the residuals from this ARMA model. Here appropriateness may be measured by the portmanteau test $\tilde{Q}_m$ or Q_m^* in section 2.3. Recall in the discussion of bilinear models some time series may appear to be white noise when only the autocorrelations are being inspected whereas the squared process could be highly correlated. Motivated by this observation McLeod and Li (1983) proposed to use the squared residual autocorrelation for diagnostic checking for possible departures from the linear ARMA model assumption. The lag-k-squared residual autocorrelation is defined by:

$$\hat{r}_{aa}(k) = \sum_{t=k+1}^{n} (\hat{a}_t^2 - \hat{\sigma}^2)(\hat{a}_{t-k}^2 - \hat{\sigma}^2) \Big/ \sum_{t=1}^{n} (\hat{\sigma}_t^2 - \hat{\sigma}^2)^2 \ , \qquad (5.7)$$

where

$$\hat{\sigma}^2 = \sum \hat{a}_t^2 / n \ .$$

For fixed M it can be shown that

$$\sqrt{n}\, \hat{\boldsymbol{r}}_{aa} = \sqrt{n}\, ((\hat{r}_{aa}(1), \ldots, \hat{r}_{aa}(M))^{\mathrm{T}} \qquad (5.8)$$

is asymptotically normally distributed as $n \to \infty$ with mean zero and unit covariance matrix. A goodness-of-fit test is provided by the portmanteau statistic

$$Q_{aa}^* = n(n+2) \sum_{i=1}^{M} \hat{r}_{aa}^2(i)/(n-i) \qquad (5.9)$$

which is asymptotically χ_M^2 distributed.

Suppose the a_t's are uncorrelated up to the fourth order moment. If there is a nonlinearity structure in X_t no ARMA model could remove all the dependence structure. In fact, an ARMA model can only remove all the second order dependence structure at best. Hence any remaining dependence on nonlinearity may be reflected by the squared residual autocorrelations (5.7). It is important to note that under the null hypothesis that the ARMA model alone is adequate Q_{aa}^* is χ_M^2 distributed asymptotically. This is different from the result in Chapter 2 where $\tilde{Q}_m$ or Q_m^* is χ_{M-p-q}^2 distributed. Many textbooks, even some very good ones, have been mistaken in stating that the number of estimated ARMA parameters have to be deducted from M in Q_{aa}^*. The rationale is that, unlike the ARMA case in (2.10), the difference between $\hat{r}_{aa}(k)$ and its population counterpart by replacing $\hat{a}_t$ with a_t, the true white noise, is only $O_p\left(\frac{1}{n}\right)$.

Intuitively this suggests that in estimating the ARMA model only information contained in the second order moments is being used and information contained in the higher order moment of X_t has not been utilized. Simulation based on an AR(1) null model in McLeod and Li (1983) suggested that the size of Q_{aa}^* when $M = 20$ is acceptable at the upper 5% level with sample size as low as 50. Q_{aa}^* can be easily computed using most statistical software routines.

Example 5.1 *We consider the Canadian Lynx data for the period 1821–1934 of Figure 5.1. The data set has been widely used as a typical nonlinear time series in the literature. Figure 5.2 gives a plot of*

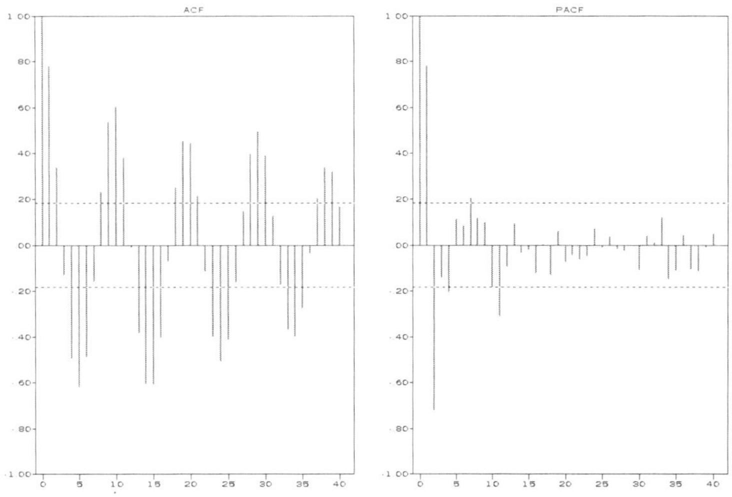

Figure 5.2 *Sample autocorrelations and partial autocorrelations of the log Lynx data*

the sample autocorrelation function (ACF) and partial ACF (PACF) of the logarithmically transformed data using the ITSM package accompanying Brockwell and Davis (1996). It can be seen that there is a cut-off after lag 11 of the PACF and this suggests that an autoregressive model of order 11 would be adequate to model the linear structure of the time series. Using the exact maximum likelihood procedure in the ITSM package to fit the model gave the following for the mean-centered log Lynx data X_t.

$$\begin{aligned} X_t &= 1.164X_{t-1} - .5397X_{t-2} + .2622X_{t-3} - .3043X_{t-4} \\ &\quad + .1457X_{t-5} - .1364X_{t-6} + 0.4811X_{t-7} - .02258X_{t-8} \\ &\quad + .1281X_{t-9} + .2092X_{t-10} - .3426X_{t-11} + a_t \end{aligned}$$

where a_t *is white noise with estimated variance* 0.1915. *The exact maximum likelihood iterative procedure converged in stable fashion after only 23 iterations. Using an* $M = 20$ *and the ITSM package the Ljung-Box statistic* Q^*_{20} *was found to have a value of* 8.1357. *This is well below the upper 5th percentile of the chi-square distribution of* $20 - 11 = 9$ *degrees of freedom which is* 16.119. *On the other hand the* $Q^*_{aa}(20)$ *statistic using the squared residual autocorrelations has a value of* 33.247. *As mentioned above, the corresponding chi-square distribution of* Q^*_{aa} *has 20 degrees of freedom (not* $20 - 11 = 9$*) with*

regression coefficient of $\hat{a}_t$ on ξ_t. Finally, let

$$F = \frac{\hat{\eta}^2(n-2M-2)}{\sum \hat{a}_t^2 - \hat{\eta}^2}. \tag{5.11}$$

Under the null hypothesis of linearity F has an asymptotic F distribution with $(1, n-2M-2)$ degrees of freedom. Note that if n is large F is χ_1^2 distributed asymptotically. The rationale of Keenan's test is that if the linear autoregression in the first step is adequate then the residual of $\hat{X}_t^2$ after removing the linear effect of $X_{t-1}, \ldots, X_{t-M}$ should have no power in explaining the residuals $\hat{a}_t$ from step 1.

Davies and Petruccelli (1986) compared the empirical size and power of Keenan's F test and the Q_{aa}^* statistic using simulation. They observed that under an AR(1) process the empirical sizes for the Q_{aa}^* statistic are satisfactory while those of F are too high if the autoregressive parameter is close to one and too low if it is close to -1. With 40 simulated series of length 100 from a threshold autoregressive model of order 1 in both regimes the F statistic detected nonlinearity in about half of the series and the Q_{aa}^* in about 1/6 of the series. However, with 160 real data series Q_{aa}^* performs slightly better (13%) than the F statistic (10%) in detecting nonlinearity. In each case an appropriate ARMA model was first fitted to the series using AIC and BIC before Q_{aa}^* was applied to the residuals.

(iii) Tsay's test

Tsay (1986) modified Keenan's F-test by including cross-product terms like $X_{t-1}X_{t-2}$ as regressors in Keenan's procedure. Specifically, let $\boldsymbol{X}_{t-1} = (1, X_{t-1}, \ldots, X_{t-p})^{\mathrm{T}}$, and let $M_{t-1} = \mathrm{vech}(\boldsymbol{X}_{t-1}\boldsymbol{X}_{t-1}^{\mathrm{T}})$ where $\mathrm{vech}(M)$ = the half-stacking vector of the matrix M on and below the main diagonal. Now consider the regression

$$X_t = \boldsymbol{X}_{t-1}^{\mathrm{T}}\boldsymbol{\phi} + M_{t-1}^{\mathrm{T}}\boldsymbol{\alpha} + e_t \tag{5.12}$$

where $\boldsymbol{\phi}$ is a $(p+1) \times 1$ vector of parameters and $\boldsymbol{\alpha}$ is a $\frac{1}{2}p(p+1) \times 1$ vector of parameters. If the linear AR(p) model is adequate in modeling X_t then $\boldsymbol{\alpha} = 0$ and the usual partial F test applies asymptotically, with degrees of freedom $\left(\frac{1}{2}p(p+1), n-p-\frac{1}{2}p(p+1)-1\right)$. Simulation in Tsay (1986) showed that the modified procedure has a larger power than the original F statistic. See also the book by Tsay (2002).

(iv) The bispectral test

The three aforementioned tests all essentially exploit on the possible nonlinear dependence structure of the time series that is reflected by the fourth order moments. The bispectral tests of Subba Rao and

Gabr (1980) are non-parametric tests making use of the third order moments of the time series. Define the quantity $C(t_1, t_2)$ by

$$C(t_1 t_2) = \mathrm{E}[(X_t - \mu)(X_{t+t_1} - \mu)(X_{t+t_2} - \mu)] \ . \qquad (5.13)$$

Here we assume that $\{X_t\}$ has finite sixth order moments and is stationary up to that order. Then the bispectral density function is just the Fourier transform of $C(t_1, t_2)$ defined by

$$\begin{aligned} f(w_1, w_2) &= \frac{1}{(2\pi)^2} \sum_{t_1=-\infty}^{\infty} \sum_{t_2=-\infty}^{\infty} C(t_1, t_2) e^{-it_1 w_1 - it_2 w_2} \ , \\ -\pi &\leq w_1 \ , \ w_2 \leq \pi \ , \end{aligned} \qquad (5.14)$$

where $i = \sqrt{-1}$. The bispectral density function is just analogous to the usual definition of the spectral density function $f(w)$ where

$$f(w) = \frac{1}{2\pi} \sum_{s=-\infty}^{\infty} \gamma(s) e^{-isw} \ , \quad -\pi \leq w \leq \pi \ , \qquad (5.15)$$

where $\gamma(s)$ is the lag s theoretical autocovariance of X_t. Given $X_1, \ldots, X_n$ the bispectral density and the spectral density can be estimated by replacing $C(t_1, t_2)$ and $\gamma(s)$ by their respective sample counterparts

$$\hat{C}(t_1, t_2) = \frac{1}{n} \sum_{t=1}^{n-l} (X_t - \bar{X})(X_{t+t_1} - \bar{X})(X_{t+t_2} - \bar{X}) \ ,$$

where $l = \max(0, t_1, t_2)$, and

$$\hat{\gamma}(s) = \frac{1}{n} \sum_{t=1}^{n-k} (X_t - \bar{X})(X_{t+k} - \bar{X}) \ .$$

Let

$$\hat{f}(w) = \frac{1}{2\pi} \sum_{l=-M}^{M} \lambda\Big(\frac{l}{M}\Big) \hat{\gamma}(s) \cos(lw) \qquad (5.16)$$

where $\lambda(\cdot)$ is a univariate lag window generator, M is a truncation point, and

$$\hat{f}(w_1, w_2) = \frac{1}{(2\pi)^2} \sum_{l_1=-M}^{M} \sum_{l_2=-M}^{M} \lambda\Big(\frac{l_1}{M}, \frac{l_2}{M}\Big) \hat{C}(l_1, l_2) e^{-il_1 w_1 - il_2 w_2}$$

where $\lambda(\cdot, \cdot)$ is a bivariate lag window generator. Choices of $\lambda(\cdot)$ could be the Parzen window

$$\lambda(l) = \begin{cases} 1 - 6l^2 + 6|l|^3 & , \quad |l| < \frac{1}{2} \\ 2(1 - |l|)^3 & , \quad \frac{1}{2} \leq l \leq 1 \\ 0 & , \quad |l| > 1 \ . \end{cases}$$

Following Subba Rao and Gabr (1980), choice of $\lambda(l_1, l_2)$ could be of the form

$$\lambda(l_1, l_2) = \lambda(l_1)\lambda(l_2)\lambda(l_1 - l_2)$$

where $\lambda(l)$ is a univariate lag window. Let

$$D_{ij} = \frac{|\hat{f}(w_i, w_j)|^2}{|\hat{f}(w_i)\hat{f}(w_j)\hat{f}(w_i + w_j)|} ,$$

where $0 < w_i < w_j < \pi$. It can be shown that if X_t is linear then

$$D_{ij} = \text{constant} \tag{5.17}$$

and a test can be based on the testing of this property. D_{ij} is approximately normally distributed by a result of Brillinger (1965). To test (5.17) a random sample of $P \times 1$ vectors $\boldsymbol{Y}_i$, $i = 1, \ldots, N$, for some N, where each $\boldsymbol{Y}_i$ has jth element D_{kl} for some integers k and l, can be formed as in Subba Rao and Gabr (1980). Let $\bar{\boldsymbol{Y}}$ be the sample mean of $\boldsymbol{Y}_i$ and $\boldsymbol{\Sigma}_Y$ be the sample covariance matrix of $\boldsymbol{Y}_1, \ldots, \boldsymbol{Y}_N$. Let $\boldsymbol{B}$ be a $(P-1) \times P$ matrix which is of the form

$$\boldsymbol{B} = \begin{bmatrix} 1 & -1 & & & \\ & 1 & \ddots & & O \\ & & \ddots & \ddots & \\ & & & \ddots & -1 \\ & O & & & 1 \end{bmatrix} \tag{5.18}$$

and

$$\boldsymbol{\beta} = \boldsymbol{B}\boldsymbol{Y} .$$

Then under the null hypothesis $\boldsymbol{\beta}$ is asymptotically Gaussian distributed with mean 0 and covariance matrix $\boldsymbol{B}\boldsymbol{\Sigma}_Y\boldsymbol{B}^{\mathrm{T}}$. Let $Q = P-1$. The test statistic (Subba Rao and Gabr, 1980) is

$$F = \frac{n-Q}{Q} T^2 \tag{5.19}$$

where

$$T^2 = n\bar{\boldsymbol{\beta}}\hat{\boldsymbol{S}}\bar{\boldsymbol{\beta}}^{\mathrm{T}}$$

where $\bar{\boldsymbol{\beta}} = \boldsymbol{B}\bar{\boldsymbol{Y}}$ and $\hat{\boldsymbol{S}} = n \cdot \boldsymbol{B}\boldsymbol{\Sigma}_Y\boldsymbol{B}^{\mathrm{T}}$. Under the null hypothesis of linearity (5.19) is F-distributed with $(Q, n-Q)$ degrees of freedom.

The test when applied to the Wölf's Annual Sunspot data and the Canadian Lynx data suggested strongly the presence of nonlinearity.

Akin to the bispectral test, Lawrance and Lewis (1987) considered the use of third order moments $\text{corr}[(X_t - \mu), \hat{a}_{t-i}^2]$ and $\text{corr}[(X_t - \mu)^2, \hat{a}_{t-i}]$ in identifying higher order dependence in certain time series. Here corr$(\cdot,\cdot)$ stands for the correlation function and $\hat{a}_t$ are residuals from a p-th order autoregression fitted to the data.

(v) Kolmogrov-Smirnov type tests

Let $\hat{a}_t$ be residuals from an autoregressive model of order p fitted to X_t. The order p can be estimated using say an information criterion such as the BIC. An and Cheng (1991) considered a Kolomogorov-Smirnov type test for linearity. Define

$$\hat{K}_{ni}(t) = \frac{1}{\sqrt{m}\hat{\sigma}} \sum_{t=p+1}^{m} \hat{a}_t I(X_{t-i} < t)$$
$$\hat{K}_{ni} = \sup_t |\hat{K}_{ni}(t)| \ ,$$

and the test statistic is

$$\hat{K}_n = \max\{\hat{K}_{ni}, i = 1, \ldots, p\} \qquad (5.20)$$

where m is an integer such that as $m \to \infty$, $m(\ln\ln(n))/n \to 0$ as $n \to \infty$. They showed that if $p = 1$, $\hat{K}_n$ converges to $K = \sup_{0\leq t} |B(t)|$ where $\{B(t)\}$ is a standard Brownian motion on $[0, 1]$. Unfortunately, when $p > 1$, the limiting distribution of the test statistic is not well established and the above limiting distribution remains ad hoc. Critical values of K can be obtained in Grenander and Rosenblatt (1957).

More recently, under a slightly different setup, Lobato (2003) defined Cramér-Von Mises and Kolmogorov-Smirnov type statistics for testing that the conditional mean of X_t is a linear autoregression of order p. He uses a sequence of alternatives that tends to the null hypothesis at a rate $n^{-1/2}$. The asymptotic distribution is found by a variant of the wild bootstrap. For details see Lobato (2003).

Koul and Stute (1999) considered a more general approach to the problem of testing the hypothesis $H_0 : E(\cdot|F_{t-1}) = \mu_t = m_t(\cdot, \theta_0)$. The proposed tests are based on a class of empirical processes marked by a function of the innovations $\psi(X_t - \mu_t)$. The choice of $\psi(\cdot)$ is up to the statistician to decide. In a related setup, Diebolt (1990) considered the model

$$X_t = T(X_{t-1}) + U(X_{t-1})a_t$$

where T and $U : R \to R$ are real continuous functions with U positive. The functions T and U are estimated non-parametrically using the regressogram approach (Tukey, 1961). Two non-parametric goodness-of-fit tests were proposed, one for T and the other for U. However, these approaches are beyond the scope of this book.

5.3 Tests for linear vs. specific nonlinear models

All the tests introduced so far can be regarded as general diagnostic tests of linearity against nonlinearity. In other words, they can be considered as pure significance tests which do not have a specific alternative in mind. Tests have been developed to test the null of linearity against alternatives of specific nonlinear models. These are usually more involved mathematically and computationally but with respect to the specific alternatives they also give higher power than pure significance tests. The first two of these are tests against the alternative of a threshold type nonlinear model, viz., threshold autoregressive models.

(i) A likelihood ratio test for threshold nonlinearity

For simplicity, we restrict the alternative threshold autoregressive model to have two regimes only. Following Chan and Tong (1990), the TAR model (5.2) with two regimes can be defined as

$$X_t - \phi_0 - \phi_1 X_{t-1} \cdots - \phi_p X_{t-p} - I(X_{t-d} \leq C)(\theta_0 + \theta_1 X_{t-1} + \cdots + \theta_q X_{t-q}) = a_t \ , \tag{5.21}$$

where $I(\cdot)$ is the indicator function and a_t is assumed to be independent and identically $N(0, \sigma^2)$ distributed. Given known d and C the null hypothesis H_0 of linearity is nested within the framework (5.21). Clearly (5.21) reduces to an AR(p) model if $\theta_0 = \theta_1 = \cdots = \theta_q = 0$. Therefore, under this situation the usual likelihood ratio test applies with the usual asymptotic chi-square distribution with q degrees of freedom. However, when C is unknown the null hypothesis is no longer nested within the alternative. Under H_0, the nuisance parameter C is absent. It is well known that under such circumstances the classical result for likelihood ratio tests is no longer true. Davies (1977, 1987) proposed that the supremum of the usual likelihood ratio test should be used in such circumstances. Let the likelihood ratio test for a particular value of C be denoted LRT(C) in (5.21). The test statistic is given by

$$\lambda = \max_{C \in \mathcal{C}} \ \text{LRT}(C) \tag{5.22}$$

where C is in a bounded subset of the real line. The asymptotic distribution of λ in general does not have a closed form. However, Chan and Tong (1990) managed to obtain tabulation results for the following two special cases:

(1) Model (5.21) takes the form,

$$X_t - \phi_d X_{t-d} - \theta_d X_{t-d} I(X_{t-d} \leq C) = a_t$$

and the null hypothesis is $H_0 : \theta_d = 0$. In this case the asymptotic null distribution of λ reduces to the distribution of

$$\sup_S \frac{B_S^2}{S - S^2} , \tag{5.23}$$

where $S = S(C) = E\{X_{t-d}^2 \mathrm{I}(X_{t-d} \leq C)\}/\mathrm{var}(X_t)$, $0 \leq S \leq 1$; $B_S = \xi_S/\sqrt{\mathrm{var}(X_t)}$, where ξ_S is a certain one-dimensional Gaussian process with zero mean. (See Chan and Tong, 1990, Appendix B.) Note that $\{B_S\}$ is a one-dimensional Brownian bridge. A Brownian bridge B_S is a Gaussian random function such the $E(B_S) = 0$ and $E(B_S B_t) = S(1-t)$ for $S \leq t$ (Billingsley, 1999). For C ranging between the 10th percentile and the 90th percentile of X_t the approximate upper 10, 5, 2.5, and 1% points for the asymptotic null distribution of S are 5.81, 7.33, 8.84, and 10.81, respectively.

(2) Model (5.21) takes the form,

$$\begin{aligned} X_t - \phi_0 - \phi_1 X_{t-1} \cdots - \phi_p X_{t-p} - I(X_{t-d} \leq C) \\ (\theta_0 + \theta_1 X_{t-1} + \cdots + \theta_p X_{t-p}) = a_t \end{aligned} \tag{5.24}$$

and $H_0 : \theta_i = 0$, $i = 0, 1, \ldots, p$.

Table 5.1 gives the critical values for λ for this case when C is within the 10th percentile and the 90th percentile of X_t. Except for the case $p = 0$ the results are just the same as those of Chan (1991). The result for case $p = 0$ is from Wong and Li (1997). Chan (1991) also gives the results when C ranges between the 25th percentile and the 75th percentiles of X_t.

The special case where no intercept terms are involved in (5.21), i.e.,

$$\begin{aligned} X_t - \phi X_{t-1} \cdots - \phi_p X_{t-p} - I(X_{t-d} \leq C)(\theta_1 X_{t-1} \\ + \cdots + \theta_p X_{t-p}) = a_t \end{aligned} \tag{5.25}$$

and $H_0 : \theta_i = 0$ $(i =, 1 \ldots, p)$ is an important case with financial time series in particular. Using simulations the approximate percentile points for the null distribution of λ are reported in Table 5.2. This table is from Wong and Li (1997). Again it is assumed that C ranges between the 10th percentile and the 90th percentile of X_t.

Chan and Tong (1990) applied the likelihood ratio test λ to both the raw Canadian Lynx data and the data after $\log_{10}$ transformation. For the raw data they used $p = 1$ and $d = 1$ and for the $\log_{10}$ transformed data they used $p = 2$ and $d = 1$. In both cases threshold nonlinearity was established. They also applied the test λ to the raw (with $p = 2$, $d = 1$) and square root transformed sunspot numbers (with $p = 2$, $d = 2$) with the same conclusion of rejecting the null hypothesis of linearity.

Table 5.1 *Upper percentage points for the asymptotic null distribution of* λ *(adapted from Chan, 1991). ©1991 The Royal Statistical Society, reproduced with the permission of Blackwell Publishing*

p	10.0%	5.0%	2.5%	1.0%
0	7.75	9.33	10.87	12.87
1	11.05	12.85	14.55	16.72
2	13.26	15.18	16.98	19.25
3	15.30	17.31	19.19	21.57
4	17.22	19.23	21.28	23.73
5	19.05	21.23	23.26	25.79
6	20.82	23.07	25.16	27.77
9	25.84	28.30	30.55	33.36
12	30.58	33.20	35.61	38.59
15	35.13	37.91	40.44	43.58
18	39.54	42.45	45.11	48.39

Table 5.2 *Upper percentage points for the asymptotic null distribution of* λ *for special case (C), the no-intercept model (Wong and Li, 1997). ©1997 Biometrika Trust, reproduced with the permission of Oxford University Press*

p	10.0%	5.0%	2.5%	1.0%
1	5.81	7.33	8.84	10.81
2	9.21	11.13	12.89	15.11
3	12.00	13.99	15.84	18.15
4	14.31	16.39	18.31	20.72
5	16.40	18.56	20.55	23.03
6	18.34	20.57	22.63	25.20
9	23.69	26.12	28.35	31.12
12	28.61	31.21	33.59	36.54
15	33.28	36.03	38.55	41.65
18	37.78	40.67	43.31	46.55

(ii) Tsay's arranged autoregression test

Tsay (1989) proposed a clever idea for testing the linear hypothesis against the alternative of threshold nonlinearity. Observe that for a sufficiently long realization $x_1, x_2, \ldots, x_n$ for the time series X_t the number of x_t's that lie in the two regimes would be nonzero. Hence, there will be some x_t's lying above C and some below. If $x_{(i)}$ denotes the i-th other statistic of x_t, that is $x_{(1)} \leq x_{(2)} \cdots \leq x_{(n)}$. Then the threshold value C must lie somewhere between the smallest observation $x_{(1)}$ and the largest observation $x_{(n)}$. In other words, there exists an integer (i_0) such that $x_{(i_0)} \leq C \leq x_{(i_0+1)}$. Let $t(j)$ be the time index corresponding, to the jth order statistic. Clearly, if $j \leq i_0$ then the observation $X_{t(j)+d}$ will be in the regime corresponding to $X_{t(j)} \leq C$. In this case $X_{t(j)+d}$ will satisfy the autoregression

$$X_{t(j)+d} = \beta_0 + \sum_{k=1}^{p} \beta_k X_{t(j)+d-k} + a_{t(j)+d} \ , \tag{5.26}$$

where $\beta_i = \phi_i + \theta_i$, if $j < i_0$.

To obtain the test we first estimate (5.26) using sufficient number of initial observations corresponding to $j = 1, \ldots, m$, where $m < i_0$. Let the predictive residuals be

$$\hat{a}_{t(m+1)+d} = X_{t(m+1)+d} - \hat{\beta}_{0,m} - \sum_{k=1}^{p} \hat{\beta}_{k,m} X_{t(m+1)+d-k}$$

and $\hat{e}_{t(m+1)+d}$ be the corresponding standardized predictive residual. We then update the regression by including the data point $X_{t(j)+d}$ in (5.26), $j = m+1$. This can be done using a recursive least squares procedure (Tsay, 1989). This procedure is repeated until all the data are included. Now consider the regression of the $\hat{e}_{t(m+j)+d}$ on $X_{t(m+j)+d-i}$, $i = 1, \ldots, p$, that is

$$\begin{aligned} \hat{e}_{t(m+j)+d} &= \alpha_0 + \sum_{i=1}^{p} \alpha_i X_{t(m+j)+d-i} + V_t \ , \\ j &= 1, \ldots, n-d-m \ , \end{aligned} \tag{5.27}$$

and compute the usual F statistic for testing $H_0 : \alpha_0$, $i = 0, \ldots, p$ in (5.26). Under the null of linearity this statistic has an asymptotic F-distribution with degrees of freedom $p+1$ and $n-d-m-p$.

The arranged autoregression can be exploited further as a tool in the identification of the threshold parameter C. See Tsay (1989). Petruccelli and Davies (1986) formed a cummulative sum (CUSUM) statistic using a similar idea. Petruccelli (1988) improved the original CUSUM test by introducing a reverse CUSUM test. Moeanaddin and Tong (1988) compared Chan and Tong's likelihood ratio test and the CUSUM tests for threshold autoregressions. Overall, they found that the likelihood ratio test performs better than the CUSUM tests.

(iii) LM tests for the bilinear model and the exponential autoregressive model

Saikkonen and Luukkonen (1988) developed a Lagrange multiplier test for the bilinear model (5.4)

$$X_t = \sum_{j=1}^{p} \phi_j X_{t-j} + \sum_{j=1}^{q} \theta_j a_{t-j} + \sum_{k=0}^{P}\sum_{l=1}^{Q} \beta_{kl} a_{t-k} X_{t-l} + a_t \ ,$$

where a_t is a Gaussian white noise process with mean 0 and variance σ^2.

Let $\boldsymbol{\theta}^{\mathrm{T}} = (\boldsymbol{\theta}_1^{\mathrm{T}}, \boldsymbol{\theta}_2^{\mathrm{T}})$ where $\boldsymbol{\theta}_1^{\mathrm{T}} = (\phi_1, \ldots, \phi_p, \theta_1, \ldots, \theta_q)$ and $\boldsymbol{\theta}_2^{\mathrm{T}} = (\beta_{01} \ldots, \beta_{PQ})$. Let $\hat{a}_t$ be the residuals from fitting the ARMA(p, q) model

$$X_t = \sum_{j=1}^{p} \phi_j X_{t-j} + \sum_{j=1}^{q} \theta_j a_{t-j} + a_t \ .$$

A Lagrange multiplier test LM1 for the presence of bilinearity can be formed by regressing $\hat{a}_t$ on the regressors $\partial a_t / \partial \boldsymbol{\theta}_1$ and $\partial a_t / \boldsymbol{\partial \theta}_2$. The partial derivatives are evaluated at $\tilde{\boldsymbol{\theta}} = (\hat{\boldsymbol{\theta}}_1, 0)$ where $\hat{\boldsymbol{\theta}}_1$ is from the fitted ARMA model above. As in (2.20), the LM1 test is given by $n \cdot R^2$ where R^2 is the coefficient of determination of the regression. Under H_0 the LM1 test has an asymptotic chi-square distribution with degrees of freedom equal to the number of terms under the double summation sign of the bilinear model (5.4).

A similar Lagrange multiplier test LM2 is also derived by the same authors for the exponential autoregressive model (Haggan and Ozaki, 1981)

$$\begin{aligned} X_t + \phi_1 X_{t-1} + \cdots + \phi_p X_{t-p} + \big\{ \exp(-\gamma X_{t-1}^2) - 1 \big\} \\ \cdot \sum_{j=1}^{p} \theta_j X_{t-j} = \mu + a_t \ . \end{aligned} \tag{5.28}$$

The null hypothesis of linearity corresponds to $H_0 : \gamma = 0$. Saikkonen, and Luukkonen (1988) compared the power of these Lagrange multiplier test with Keenan's and the McLeod-Li tests.

(iv) Tests for smooth transition threshold autoregressive models

The threshold model (5.2) exhibits an abrupt change in regime depending on when the X_{t-d} will cross the threshold value C. In reality this need not be so and changes can be smooth. To cater for this possibility Chan and Tong (1986) first considered the smooth transition threshold model using an S-shaped function to model the transition from one regime to the other. Luukkonen, Saikkonen and Teräsvirta

(1988) considered testing linearity against smooth transition autoregressive models. A smooth transition autoregressive (STAR) model can be defined as,

$$X_t = \phi_0 + \boldsymbol{\phi}^{\mathrm{T}} \boldsymbol{X}_{t-1} + (\theta_0 + \boldsymbol{\theta}^{\mathrm{T}} \boldsymbol{X}_{t-1}) F(z_t) + a_t \ , \qquad (5.29)$$

where $\boldsymbol{X}_{t-1} = (X_{t-1}, \ldots, X_{t-p})^{\mathrm{T}}$, $\boldsymbol{\phi} = (\phi_1, \ldots, \phi_p)^{\mathrm{T}}$, $\boldsymbol{\theta} = (\theta_1, \ldots, \phi_p)^{\mathrm{T}}$, $z_t = \gamma(a^{\mathrm{T}} \boldsymbol{X}_{t-1} - C)$, $\gamma > 0$, $\boldsymbol{a} = (a_1, \ldots, a_p)^{\mathrm{T}}$. The function $F(\cdot)$ has an S-shaped and continuous graph. Examples of $F(\cdot)$ include any cumulative distribution function. Two examples are the cumulative distribution functions of the standard normal distribution $\Phi(\cdot)$ and the logistic function $F(z) = e^z/(1+e^z)$. Under the null hypothesis of linearity $\theta_0 = \theta_1 = \cdots \theta_p = 0$ and X_t is just an AR(p) process. Note that if γ tends to infinity $F(\cdot)$ tends to one and this gives the original threshold autoregressive model. Estimation of (5.29) can be done by means of the maximum likelihood approach. Chan and Tong (1986) derived the asymptotic distribution of the maximum likelihood estimates. Luukkonen et al. (1988) proposed several tests for linearity against smooth transition autoregressive models. The first test is by replacing $F(z)$ with a first order Taylor approximation of $F(z)$ around $z = 0$. That is

$$F(z) \cong T(z) = g_1 z$$

where $g_1 = dF(z)/dz\big|_{z=0}$. In this case (5.29) reduces to a linear model

$$X_t = \phi_0 + \boldsymbol{\phi}^{\mathrm{T}} \boldsymbol{X}_{t-1} + \pi_0(\boldsymbol{a}^{\mathrm{T}} \boldsymbol{X}_{t-1} - C) + \boldsymbol{\pi}^{\mathrm{T}} \boldsymbol{X}_{t-1}(\boldsymbol{a}^{\mathrm{T}} \boldsymbol{X}_{t-1} - C) + a_t \qquad (5.30)$$

where $\pi_0 = \gamma g_1 \theta_0$ and $\boldsymbol{\phi} = \gamma g_1 \boldsymbol{\theta}_1$. Under H_0, $\pi_i = 0$, $i = 0, 1, \ldots, p$. Since C is unknown it is necessary to reparameterize (5.30) before we can have a meaningful test. Multiplying out (5.30), after some algebra (5.30) can be written as

$$X_t = \alpha_0 + \boldsymbol{\alpha}^{\mathrm{T}} \boldsymbol{X}_{t-1} + \sum_{i=1}^{p} \sum_{j=1}^{p} \phi_{ij} X_{t-i} X_{t-j} + a_t \ , \qquad (5.31)$$

for some parameters α_0, $\boldsymbol{\alpha}$ and ϕ_{ij}. The test now becomes a test of $H_0'' : \phi_{ij} = 0$, $i = 1, \ldots, p, j = 1, \ldots, p$. The classical F statistic can be applied to (5.21) which has an asymptotical $\chi^2_{\frac{1}{2}p(p+1)}$ distribution. However, the ϕ_{ij}'s do not involve the θ_S and this may result in a low power for the F test. To overcome this deficiency, Luukkonen et al. (1988) considered also a third order approximation of $F(z)$ by the function

$$T_3(z) = g_1 z + g_3 z^2 \qquad (5.32)$$

where

$$g_3 = \frac{1}{6}\left(\left. \frac{d^3 F(z)}{dz^3} \right|_{z=0} \right) .$$

The third test is a modification of the first order test by including the terms X_{t-j}^3, $j = 1, \ldots, p$. Some simulation experiments suggested that the third order test is the most powerful of the three. See Luukkonen et al. (1988) and Granger and Teräsvirta (1993).

5.4 Goodness-of-fit tests for nonlinear time series

It would be very useful for the statistician fitting nonlinear time series models if there existed some general goodness-of-fit tests for such models. In the same spirit as with ARMA models it is is reasonable to regard that a nonlinear time series model is a good fit to the data if its residual autocorrelations are approximately zero. The asymptotic distribution of residual autocorrelations for a general stationary nonlinear time series has been derived by Li (1992). A generalization to nonlinear models with random coefficients is obtained by Hwang, Basawa, and Reeves (1994). Hwang et al. (1994) proposed also a goodness-of-fit test based on the prediction errors. We consider first the results of Li (1992) and use the same notation $\hat{a}_t$ for the residuals and $\hat{r}_k$ for the lag k residual autocorrelation which is also defined similarly as in (2.4).

Assume that $\{X_t\}$ satisfies the nonlinear model

$$X_t = f(F_{t-1}; \boldsymbol{\phi}) + a_t \ , \qquad (5.33)$$

where f is a known nonlinear function of past X_t's and $\boldsymbol{\phi}$ is a $p \times 1$ vector of parameters. Let $\{X_t\}$ be a stationary and ergodic time series, with F_t the σ-field generated by $\{X_t, X_{t-1}, \ldots\}$. The function f is assumed to have continuous second order derivatives almost surely. The noise process $\{a_t\}$ is assumed to be independent, with mean zero, variance σ_a^2, and finite fourth order moment. It is further assumed that (5.33) is invertible or equivalently $\{a_t\}$ is measurable with respect to F_t.

Let the length of realization be n. Let the lag k white noise autocovariance be $C_k = \sum a_t a_{t-k}/n$ $(k = 1, \ldots, M)$ and let $r_k = C_k/C_0$, $\boldsymbol{r} = (r_1, \ldots, r_M)^{\mathrm{T}}$. Denote by $\hat{C}_k$ the corresponding residual autocovariances obtained by replacing a_t in C_k with the residuals $\hat{a}_t$. The residuals $\{\hat{a}_t\}$ are assumed to be from a least squares fit of (5.33) to $\{X_t\}$. Define the lag k residual autocorrelations to be $\hat{r}_k = \hat{C}_k/\hat{C}_0$. Using a Taylor series expansion of $\hat{r}_k$ it can be shown that the asymptotic distribution of $\hat{r}_k$ does not depend on $\hat{C}_0$ and therefore we can ignore $\hat{C}_0$ in deriving the asymptotic distribution of $\hat{r}_k$. The result for $\hat{r}_k$ will follow from that of $\hat{C}_k$ by scaling.

Let $\hat{\boldsymbol{r}} = (\hat{r}_1, \ldots, \hat{r}_M)^{\mathrm{T}}$. The residual variance $\hat{\sigma}_a^2$ is estimated by $\hat{C}_0$. If $\{a_t\}$ have finite fourth order moments, then it is well known that $r\sqrt{n}$ is asymptotically normally distributed with mean zero and variance $\mathbf{1}_M$, where $\mathbf{1}_M$ is the $M \times M$ identity matrix. Under regularity conditions as given by Klimko and Nelson (1978) the least squares estimator $\hat{\boldsymbol{\phi}}$ of $\boldsymbol{\phi}$ can be shown to be asymptotically normally distributed with mean $\boldsymbol{\phi}$ and covariance matrix $\sigma_a^2 V^{-1}/n$, where

$$\boldsymbol{V} = E\Big\{n^{-1}\sum(\partial a_t/\partial\boldsymbol{\phi})(\partial a_t/\partial\boldsymbol{\phi})^{\mathrm{T}}\Big\}\,.$$

Denote $f(F_{t-1}, \boldsymbol{\phi})$ by f_{t-1}. Suppose that $E(\partial f_{t-1}/\partial\boldsymbol{\phi} a_{t-j})$ exists for $j = 1, \ldots, M$, and that corresponding sample averages converge in probability to the respective expected values. A sufficient condition for the latter would be that the covariance between $a_{t-j}\partial f_{t-1}/\partial\boldsymbol{\phi}$ and $a_{t'-j}\partial f_{t'-1}/\partial\boldsymbol{\phi}$ tends to zero as $|t - t'| \to \infty$. This seems to be a reasonable assumption in practice. The next two lemmas follow using Taylor series expansion of $\sum a_t^2$ and $\hat{C}_k$.

Lemma 5.1 *The asymptotic cross-covariance between* $\sqrt{n}(\hat{\boldsymbol{\phi}} - \boldsymbol{\phi})$ *and* $\sqrt{n}\boldsymbol{C} = \sqrt{n}(C_1, \ldots, C_M)^{\mathrm{T}}$ *is equal to* $\sigma_a^2 V^{-1} J$*, where*

$$J = E\Big\{\sum \partial f_{t-1}/\partial\boldsymbol{\phi} a_{t-1}, \ldots, \sum \partial f_{t-1}/\partial\boldsymbol{\phi} a_{t-M}\Big\} n^{-1}\,.$$

Proof. This follows from the standard result

$$\hat{\boldsymbol{\phi}} - \boldsymbol{\phi} \sim \Big(\sum \partial f_{t-1}/\partial\boldsymbol{\phi}\,\partial f_{t-1}^{\mathrm{T}}/\partial\boldsymbol{\phi}\Big)^{-1}\Big(\sum \partial f_{t-1}/\partial\boldsymbol{\phi} a_t\Big)\,.$$

Lemma 5.2 *For large* n*,* $\hat{\boldsymbol{C}} \sim \boldsymbol{C} - \boldsymbol{J}^{\mathrm{T}}(\hat{\boldsymbol{\phi}} - \boldsymbol{\phi})$.

Proof. This follows from a Taylor series expansion of $\hat{C}_k$ about $\boldsymbol{\phi}$ and evaluated at $\hat{\boldsymbol{\phi}}$. □

From these two lemmas and the martingale central limit theorem (Billingsley, 1961) we have the following theorem of Li (1992).

Theorem 5.1 *The large sample distribution of* $\hat{\boldsymbol{r}}\sqrt{n}$ *is normal with mean zero and covariance matrix* $\mathbf{1}_M - \sigma_a^{-2}\boldsymbol{J}^{\mathrm{T}}\boldsymbol{V}^{-1}\boldsymbol{J}$.

Note that, for autoregressive moving average models, $\boldsymbol{V}$ and $\boldsymbol{J}$ can be evaluated in terms of $\boldsymbol{\phi}$. For nonlinear models closed form expressions for these quantities are usually unavailable. Our proposal here is to use observed quantities instead of the expectations. This is in some sense analogous to the use of observed rather than expected Fisher information (Efron and Hinkley, 1978). The theorem suggests that we can use the statistic

$$Q(M) = n \cdot \hat{\boldsymbol{r}}^{\mathrm{T}}(\mathbf{1}_M - \sigma_a^{-2}\boldsymbol{J}^{\mathrm{T}}\boldsymbol{V}^{-1}\boldsymbol{J})^{-1}\hat{\boldsymbol{r}} \tag{5.34}$$

as a general goodness-of-fit test for model (5.33). $Q(M)$ has an asymptotic chi-squared distribution of M degrees of freedom if (5.33) is an adequate model.

A small simulation experiment was conducted in Li (1992) to compare the asymptotic and the empirical standard errors of $\hat{r}_k$ in threshold models. The design of the experiment was as follows. We considered a simple TAR (2; 1, 1) model $X_t = \phi_1 X_{t-1} + a_t$ if $X_{t-1} > 0$; and $X_t = \phi_1' X_{t-1} + a_t$ otherwise, where $\{a_t\}$ were normally distributed with mean 0 and variance 1. Then it can be easily shown that $\boldsymbol{V} \simeq n^{-1}(\boldsymbol{X}^{\mathrm{T}}\boldsymbol{X})$, where $\boldsymbol{X}$ is given by Tong (1983, p.140). Similarly, elements of $\boldsymbol{J}$ can be shown to be the limits in probability of the quantities $\sum X_{t-1}a_{t-k}I_j/n$, where $k = 1, \ldots, M$, $j = 1, 2$. Here I_1 indicates $X_{t-1} > 0$ and $I_2 = 1 - I_1$. For each pair (ϕ_1, ϕ_1'), 1000 independent realizations each of length 200 were generated. The values of (ϕ_1, ϕ_1') considered were $(0.5, -0.5)$, $(-0.8, 0.8)$, $(0.95, -0.95)$, $(0.8, 0.3)$, $(-0.8, -0.3)$. The series were generated and fitted using IMSL subroutines. The sample variances $V(\hat{r}_k)$ of $\hat{r}_k$ over the 1000 replications were computed for each model. Denote $\sqrt{V(\hat{r}_k)}$ by Sd_k. These were taken to be the "true" standard errors of $\hat{r}_k$. The asymptotic variances $C(\hat{r}_k)$ were also estimated for each realization using Theorem 5.1. The sample averages of $C(\hat{r}_k)$ were obtained and were denoted as $\bar{C}_k$. The results for $\sqrt{\bar{C}_k}$ and Sd_k $(k = 1, \ldots, 6)$ are reported in Table 5.3.

As in the linear autoregressive situation the results in Table 5.3 showed that the "true" standard errors for $\hat{r}_k$ could be smaller than the value $1/\sqrt{200} = 0.0707$. This discrepancy is more prominent if the values of k are small. Consequently, using $1.96/\sqrt{n}$ as critical value would give a very conservative confidence limit for the first few residual autocorrelations. Note also the much closer match between $\sqrt{\bar{C}_k}$ and Sd_k. This suggests that the result could be usefully applied to give more accurate standard errors in practice resulting in a more stringent criterion in diagnostic checking for threshold models. This is also consistent with the observations made in earlier chapters that the first few $\hat{r}_k$ should be given more careful scrutiny. Note that as k becomes larger both Sd_k and $\sqrt{\bar{C}_k}$ approach the value $1/\sqrt{n}$.

Hwang, Basawa, and Reeves (1994) extended Li's result to include linear and nonlinear models with random parameters. They considered the

Table 5.3 *Empirical results for residual autocorrelations in* $TAR(2;1,1)$ *models,* $n = 200$*, 1000 replications (Li, 1992).* ©*1992 Biometrika Trust, reproduced with the permission of Oxford University Press*

ϕ_1, ϕ_1'		$k=1$	$k=2$	$k=3$	$k=4$	$k=5$	$k=6$
$(0.5, -0.5)$	Sd_k	0.0282	0.0703	0.0674	0.0719	0.0709	0.0706
	$\sqrt{\bar{C}_k}$	0.0277	0.0698	0.0704	0.0704	0.0704	0.0704
$(-0.8, 0.8)$	Sd_k	0.0489	0.0688	0.0663	0.0719	0.0709	0.0706
	$\sqrt{\bar{C}_k}$	0.0477	0.0675	0.0695	0.0701	0.0703	0.0704
$(0.95, -0.95)$	Sd_k	0.0626	0.0672	0.0660	0.0714	0.0704	0.0702
	$\sqrt{\bar{C}_k}$	0.0601	0.0679	0.0689	0.0694	0.0697	0.0699
$(0.8, 0.3)$	Sd_k	0.0475	0.0630	0.0653	0.0711	0.0704	0.0701
	$\sqrt{\bar{C}_k}$	0.0459	0.0636	0.0678	0.0693	0.0699	0.0702
$(-0.8, -0.3)$	Sd_k	0.0385	0.0637	0.0659	0.0719	0.0706	0.0705
	$\sqrt{\bar{C}_k}$	0.0376	0.0632	0.0689	0.0700	0.0704	0.0704

following p-th order nonlinear autoregression

$$X_t = H(\boldsymbol{X}_{t-1}, \boldsymbol{Z}_t; \boldsymbol{\phi}) + a_t \tag{5.35}$$

where $\{a_t\}$ is a sequence of i.i.d random errors with mean 0 and variance σ_a^2, $\boldsymbol{X}_{t-1} = (X_{t-1}, \ldots, X_{t-p})^{\mathrm{T}}$ and $\boldsymbol{\phi}$ is a $p \times 1$ vector of parameters. The sequence of random vectors $\mathbf{Z}_t$ are unobservable and are assumed to be i.i.d with mean zero and independent of $\{a_t\}$. The model (5.35) includes both linear and nonlinear models with possible random coefficients. For example, the random coefficient autoregressive (RCA) model (Nicholls and Quinn, 1982):

$$X_t = (\phi_1 + Z_{t1})X_{t-1} + \cdots + (\phi_p + Z_{tp})X_{tp} + a_t \ .$$

Similarly we can define a random coefficient threshold autoregressive model of order one:

$$X_t = (\phi_1 + Z_{t1})X_{t-1} + a_t \ , \qquad \text{if} \quad X_{t-1} > C$$

$$X_t = (\phi_1' + Z_{t1}')X_{t-1} + a_t' \ , \qquad \text{otherwise} \ ,$$

where Z_{t1} and Z_{t1}' are i.i.d. sequences of random variables with mean zero. The sequences $\{Z_{t1}\}$ and $\{Z_{t1}'\}$ are also assumed to be independent of each other. Other parameters are defined as in (5.1) and extensions to higher order threshold autoregressions are direct.

Let

$$\begin{aligned} M(\boldsymbol{X}_{t-1};\boldsymbol{\phi}) &= E_{\boldsymbol{\phi}}(X_t|F_{t-1}) \\ &= E_{\boldsymbol{\phi}}\big(H(\boldsymbol{X}_{t-1},\boldsymbol{Z}_t;\boldsymbol{\phi}|F_{t-1}\big) , \end{aligned}$$

where F_{t-1} is the information contained in the past X_t's up to time $t-1$.

Given a realization of X_t with length n we can estimate $\boldsymbol{\phi}$ as before using conditional least squares. Let ∇M_{t-1} be the $p \times 1$ vector of the partial derivatives of $M(\boldsymbol{X}_{t-1};\boldsymbol{\phi})$ with respect to $\boldsymbol{\phi}$. Denote the estimate by $\hat{\boldsymbol{\phi}}$. Let $a_t = a_t(\boldsymbol{\phi}) = X_t - M(\boldsymbol{X}_{t-1};\boldsymbol{\phi})$ and let the residuals $\hat{a}_t = a_t(\hat{\boldsymbol{\phi}}) = X_t - M(\boldsymbol{X}_{t-1};\phi)$. Define the residual autocorrelations $\hat{r}_k$ as before and let $\boldsymbol{r} = (r_1,\cdots,r_M)^{\mathrm{T}}$ for some M. Let

$$\boldsymbol{V} = E_{\boldsymbol{\phi}}[\nabla M_{t-1}\cdot\nabla M_{t-1}^{\mathrm{T}}] ,$$

and

$$m_i = E[a_t(\boldsymbol{\phi})\nabla M_{t-i-1}] .$$

The following theorem gives an extension of Theorem 5.1 (Hwang et al., 1994).

Theorem 5.2 *Under regularity conditions mentioned in the paragraph defining (5.33)*

$$\sqrt{n}(\hat{r}_1,\ldots,\hat{r}_M) \xrightarrow{d} N_M(0,\boldsymbol{\Sigma})$$

where $\xrightarrow{d}$ *denotes convergence in distribution,* $\boldsymbol{\Sigma}$ *is the* $M \times M$ *matrix with* (i,j)*th element,*

$$\Sigma_{ij} = \sigma_a^{-4}E[a_t^2\{a_{t-i} - m_i^{\mathrm{T}}\boldsymbol{V}^{-1}\cdot\nabla M_{t+1}\}\cdot\{a_{t-j} - m_j^{\mathrm{T}}\nabla M_{t-1}\}]$$

where $\sigma_a^2 = E(a_t^2)$.

Based on Theorem 5.2 the portmanteau test $Q(M)$ (5.34) can also be used for time series models with random coefficients. In this case, $\mathbf{1}_M - \sigma_a^{-2}\boldsymbol{J}^{\mathrm{T}}\boldsymbol{V}\boldsymbol{J}$ is replaced by $\boldsymbol{\Sigma}$ above. In case that $\boldsymbol{\Sigma}$ is singular we can replace it by $\boldsymbol{\Sigma}^-$ a generalized inverse of $\boldsymbol{\Sigma}$ and

$$Q(M) = \boldsymbol{r}^{\mathrm{T}}\boldsymbol{\Sigma}^-\boldsymbol{r} \xrightarrow{d} \chi_M^2 \tag{5.36}$$

where $r = \mathrm{rank}(\boldsymbol{\Sigma})$.

As in Li (1992), Hwang et al. (1994) observed that the large sample variance of $\sqrt{n}\hat{r}_k$ is close to one and they proposed the use of the statistic

$$D_n(M) = n\cdot\boldsymbol{r}^{\mathrm{T}}\boldsymbol{r} \tag{5.37}$$

and treated $D_n(M)$ as asymptotically χ_{M-p}^2 distributed if the model is adequate. How good the approximation is, however, depends on both the model and sample size. The author of this book would like to suggest the use of

$$\tilde{D}_n(M) = n \cdot \sum_{i=p+1}^{M} \hat{r}_i^2 \tag{5.38}$$

which is better approximated by a χ^2_{M-p} distribution in large samples than $D_n(M)$.

Hwang et al. (1994) further proposed a goodness-of-fit test based on the prediction errors. Let the data be denoted by $X_1, \ldots X_n, X_{n+1} \ldots, X_{n+k}$. Pretend that $X_{n+1} \cdots X_{n+k}$ are unknown. Let the one-step ahead prediction of X_{n+i} given F_{n+i-1} be

$$\hat{X}_{n+i} = E_{\boldsymbol{\phi}}(X_{n+i}|F_{n+i-1}) \ .$$

Let the prediction errors be

$$e_{n+i}(\boldsymbol{\phi}) = X_{n+i} - \hat{X}_{n+i} \ .$$

Let

$$R_{n+i} = e_{n+i}(\hat{\boldsymbol{\phi}})$$

where $\hat{\boldsymbol{\phi}}$ is the conditional least squares estimate of $\boldsymbol{\phi}$. Let τ^2_{n+i} be the corresponding one-step ahead prediction variance. Expressions for the τ^2_{n+i} will be model dependent. In the special case of (5.33) this is just σ^2_a. However, for random coefficient models this will depend on i. For example, for a random coefficient, autoregressive model of order p,

$$\tau^2_{n+i} = \sigma^2_a + \sigma^2_z \sum_{j=1}^{p} X_{n+i-j} \ .$$

Then the statistic

$$W(n) = \sum_{i=1}^{k} R^2_{n+i} \tau^{-2}_{n+i} \tag{5.39}$$

has an asymptotic χ^2_k distribution under the null hypothesis that the model is adequate. A small simulation in Hwang et al. (1994) suggested that a sample size of 400 or more may be needed to give an accurate approximation to the null distribution.

5.5 Choosing two different families of nonlinear models

In recent years there has been rapid growth in the literature on nonlinear time series models. Many different types of models have been suggested. As mentioned in the beginning of this chapter, two major classes of nonlinear models are the threshold models (Tong (1978), Tong and Lim

(1980)) and the bilinear models (Granger and Andersen (1978), Subba Rao (1981)). The recent book by Tong (1990) contains a comprehensive summary of most of the proposed nonlinear models. A natural and important problem is to develop tests to discriminate among the various models. Many tests have been proposed for testing different nonlinear models against linear (ARMA) models but not among nonlinear models. Saikkonen and Luukkonen (1988) gave a summary review of the former procedures. For the latter, various informal arguments have been suggested. For example, it has been argued that threshold models can mimic limit cycle behavior but bilinear models cannot (Tong and Lim (1980)). Consequently, one should consider threshold models for data that appear to have a limit cycle. Another common approach is to compare the post sample forecast ability of the different models (Ghaddar and Tong (1981)) or the residual sum of squares (Gabr and Subba Rao (1981)). Other arguments include parsimony in terms of model parameters and whiteness of residuals. Although these arguments are valid and important it may still be beneficial if formal tests can be developed for distinguishing between different nonlinear models. Clearly, the problem is more difficult than testing nonlinearity vs. linearity since different types of nonlinear models in general cannot be nested within one another. Under the assumption of Gaussian innovations and nested models, comparing residual sums of squares is equivalent to the likelihood ratio test which is, in general, asymptotically chi-squared distributed under the null hypothesis. However, for non-nested models the likelihood ratio statistic will not normally have an asymptotic chi-squared distribution and thus the comparison of residual variances does not usually fit into the hypothesis testing framework. A possible approach is to consider a Cox test for separate families of hypotheses (Cox (1962)). This, however, requires evaluating the expectation and variance of the log-likelihood ratio under the null hypothesis. For nonlinear time series this is a difficult task. Li (1989) proposed a bootstrap procedure to overcome this difficulty. Earlier Williams (1970) and Aguirre-Torres and Gallant (1982) have applied a similar approach in a non-time series context. Wahrendorf, Becher, and Brown (1987) consider a related methodology in survival studies.

5.5.1 The bootstrapped Cox-test

Let $\boldsymbol{X} = (X_1, \ldots, X_n)^{\mathrm{T}}$ be a random vector. Suppose that under the null hypothesis H_o the probability density function is $f(\boldsymbol{X}, \gamma)$ where γ is an unknown vector parameter. Suppose that under the alternative H_A, the probability density function is $g(\boldsymbol{X}, \beta)$, where β is again an unknown vector parameter. Suppose that f and g belong to separate families. Let $\hat{\gamma}$ and $\hat{\beta}$ be maximum likelihood estimates of γ and β under H_o and H_A, respectively. Denote by $L_f(\hat{\gamma})$ and $L_g(\hat{\beta})$ the corresponding maximized log-likelihood functions. Cox (1962) proposed the test statistic

$$T_f = L_f(\hat{\gamma}) - L_g(\hat{\beta}) - E_{\hat{\gamma}}\{L_f(\hat{\gamma}) - L_g(\hat{\beta})\} \tag{5.40}$$

where $E_{\hat{\gamma}}$ denotes expectation under H_o. For independent X_i's, Cox (1962) showed that T_f is under certain regularity conditions asymptotically normally distributed under H_o. It is not difficult to conjecture that a similar result will hold for dependent X_i's provided that certain mixing or martingale type conditions are satisfied. Indeed Guégan (1981) considered one such generalization and applied her method to stationary ARMA processes. However, in many situations it is the evaluation of $E_{\hat{\gamma}}(L_f(\hat{\gamma}) - L_g(\hat{\beta}))$ and the corresponding variance that present the greatest difficulties. Furthermore, the asymptotic normal distribution may be different from the finite sample distribution. Thus we propose to approximate the finite sample distribution of T_f using the parametric bootstrap method (Efron, 1982). Our procedure can be stated as follows.

Step (1). Given a realization $\{x_1, \ldots, x_n\}$ of the time series we find the best fitting models under the two separate families of models. Denote these two models by M_o and M_A, respectively corresponding to H_o and H_A.

Step (2). For a large enough positive integer B, B sets of artificial realizations $R_k = \{x_{1k}^*, \ldots, x_{nk}^*\}$, $1 \leq k \leq B$, are generated under M_o. Maximum likelihood estimates $\hat{\gamma}_k^*$ and $\hat{\beta}_k^*$ are then obtained for each of the realizations. An approximation to the distribution of $C_f = L_f(\hat{\gamma}) - L_g(\hat{\beta})$ under H_o can now be obtained from the empirical distribution of $C_{fk}^* = L_f(\hat{\gamma}_k^*) - L_g(\hat{\beta}_k^*)$.

Step (3). The hypothesis H_o is rejected at level α if $C_f = L_f(\hat{\gamma}) - L_g(\hat{\beta})$ exceeds the $[B\alpha]$ order statistic of C_{fk}^*.

In the next section we will see how this procedure can be applied to distinguish some simple bilinear and threshold models. An example based on the Wölf sunspot numbers is also considered.

Some simulation experiments were conducted to study the effectiveness of the proposed Cox test in discriminating between simple bilinear and threshold models. In the first experiment realizations of the bilinear model (M_o)

$$X_t = a\, X_{t-1} + b\, X_{t-k}\, e_{t-l} + e_t\ , \tag{5.41}$$

where $\{e_t\}$ were Gaussian with mean zero and variance one, were generated. The methodology proposed above was first applied with H_o given by (5.41). The alternative H_A was a threshold model (M_A)

$$\begin{aligned} X_t &= \phi_1 X_{t-1} + a_t, \quad &\text{if } X_{t-1} \geq 0 \ , \\ &= \phi_2 X_{t-2} + a_t, \quad &\text{otherwise} \ , \end{aligned} \tag{5.42}$$

where $\{a_t\}$ were assumed to be normally distributed. It was assumed that the only unknown parameters were either (a, b) or (ϕ_1, ϕ_2). Note that apart from a scaling constant $C_f \cong n\big(\log \hat{\sigma}_e^2/\hat{\sigma}_a^2\big)$ where n is the length of realization and $\hat{\sigma}_e^2$, $\hat{\sigma}_a^2$ are the residual variance of $\hat{e}_t$ and $\hat{a}_t$, respectively. In Step 2, the realization $\{x_{tk}^*\}$ were generated by resampling with replacement from the empirical distribution of $\hat{e}_t$. Depending on values of k and l, the observations x_1 and/or x_2 were considered as fixed. Alternatively, one may also consider sampling from the normal distribution $N\big(0, \hat{\sigma}_e^2\big)$. However, after a few experiments were performed using this latter approach, it was observed that the test appeared to have less power than the procedure adopted here. Empirical significance levels when $\gamma = 0.05$ and 0.10 are reported in Table 5.4. In addition, the empirical significance levels of the standardized Cox test at the upper 0.05 level based on the asymptotic normal distribution are also reported. Both n and B were chosen to be 100. In practice, a larger B value in the range $(200, 400)$ is preferable. The parameter value of (a, b) is $(0.5, 0.2)$ and $(k, l) = (1, 2)$, $(1, 1)$, and $(2, 1)$. There were 100 independent replications for each combination of k and l. The white noise series was generated by the IMSL subroutine DRNOA. Subroutines DLSQRR and DBCONF were used to estimate the threshold model and the bilinear models, respectively. The bootstrap sampling step was performed using the IMSL subroutine RNUND. The empirical power of the Cox test was also considered. In this case, the null hypothesis is the threshold model (5.42) and the alternative is the bilinear model (5.41). Here $\{x_{tk}^*\}$ were generated using $\hat{\phi}_1, \hat{\phi}_2$ and the empirical distribution of $\hat{a}_t$. The results are also reported in Table 5.4.

In the second experiment, realizations from the threshold model (5.42) were generated. The white noise process a_t has mean zero and variance one. The values of ϕ_1 and ϕ_2 are 0.8 and 0.0, respectively. Three bilinear alternatives were entertained with $(k, l) = (1, 2)$, $(1, 1)$, and $(2, 1)$, respectively. Other parameters remain unchanged as in the previous experiment. The results are also reported in Table 5.4.

It can be seen from Table 5.4 that the significance levels of the bootstrapped Cox test are in general somewhat different from their expected values. Nevertheless, the very first case of experiment (I) gave results that are very close to the expected. In most of the other cases the empirical significance levels appear to lean toward smaller values. This may be due partly to small n and B values and clearly further simulations are needed here. The results on the power of the Cox test are more encouraging. In all cases, the power of the test is at least around 0.40 and sometimes much higher. This suggests that reasonable discriminating power may still be obtainable although the test may be a conservative one.

Table 5.4 *Empirical significance levels and power of C_f and the standardized Cox (Std) tests*

		Experiment I. True Model: a Bilinear Model					
		H_o			H_o		
		Bilinear Model (5.41)			Threshold Model (5.42)		
		$\alpha = 0.10$	0.05	Std. test (0.05)	0.10	0.05	Std. test (0.05)
K	l						
1,	2	0.09	0.05	0.04	0.89	0.77	0.58
1,	1	0.19	0.09	0.07	0.76	0.66	0.55
2,	1	0.05	0.02	0.02	0.61	0.54	0.43
		Experiment II. True Model: a Threshold Model					
		H_o			H_o		
		Threshold Model (5.42)			Bilinear Model (5.41)		
		$\alpha = .10$	0.05	Std. test (0.05)	0.10	0.05	Std. test (0.05)
K	l						
1,	2	.03	0.01	0.01	0.79	0.55	0.65
1,	1	.03	0.01	0.01	0.48	0.37	0.40
2,	1	.03	0.01	0.01	0.85	0.68	0.76

Example 5.3 *The Wölf annual sunspot numbers are considered as a real example. Tong and Lim (1980) considered a $SETAR(2;4;12)$ model while Gabr and Subba Rao (1981) suggested that a subset $BL(9,0,8,6)$ model gave a better fit. The following simplification of the Gabr and Suba Rao model is considered as the true model in our study,*

$$X_t - a_1 X_{t-1} - a_2 X_{t-2} - a_3 X_{t-9} - b_1 X_{t-2} e_{t-1} = \mu + e_t \, .$$

Note that the bilinear term considered here corresponds to the one with the largest coefficient in Gabr and Subba Rao (1981, eqn(5.3)). The alternative is the $SETAR(2;4;12)$ model,

$$X_t = \mu_1 + \phi_1 X_{t-1} + \phi_2 X_{t-2} + \phi_3 X_{t-3} + \phi_4 X_{t-4} + a_t, \quad \text{if} \quad X_{t-3} \leq 36.6$$

$$= \mu_2 + \sum_{i=1}^{12} \phi'_i X_{t-i} + a_t \ , \qquad\qquad otherwise \ .$$

The only difference from Tong and Lim is that here a_t is assumed to have the same variance in both equations. As in Gabr and Subba Rao (1981) we considered the observations from 1700–1920. Again the value of B was taken to be 100. The estimated model parameters were $(\hat{a}_1, \hat{a}_2, \hat{a}_3, \hat{b}_1, \hat{\mu})$ $= (1,322, -.6329, 0.1253, 0.0041, 8.3192)$, $(\hat{\phi}_1, \hat{\phi}_2, \hat{\phi}_3, \hat{\phi}_4, \hat{\mu}) = (1.7046,$ $-1.1656, 0.2261, 0.1738, 9.6846)$, $\hat{\mu}_2 = 7.8851$ and $\hat{\phi}'_1$ to $\hat{\phi}'_{12}$ were 0.7679, −0.0750, −0.1775, 0.1618, −0.2263, 0.0270, 0.1537, −0.2616, 0.3374, −0.4123, 0.4492 and −0.0509, respectively. Using the bootstrap, the upper 0.05 and 0.10 critical values for C_f under the bilinear model were found to be 34.42 and 28.36, respectively. The value of the statistic was 51.35. At the same time the standardized statistic had a value of 3.52. Assuming normal theory this value had a p-value of 0.0002. For the comparison to be a fair one, the roles of the null and alternative hypotheses were reversed and the same bootstrap procedure was repeated. The lower 0.05 and 0.10 critical values of C_f under the threshold model were found to be 88.50 and 67.50, respectively. The standardized statistic had a value of 2.55 which has a p-value of 0.0055. Thus, based on the tests both threshold and bilinear models were rejected. In fact, the value of C_f was just about mid-way between the two 5% critical values. On the other hand, the p-values do suggest that there may be somewhat more evidence for the threshold model. Perhaps the truth is somewhere in between? One reservation of the above approach is that the original subset bilinear model was not used owing to numerical difficulties. On the other hand, we had also assumed the residuals for the two branches of the threshold model to be the same.

From the simulation experiment and the example, it seems that a Cox test based on the bootstrap methodology is a rather feasible tool in discriminating nonlinear models. A major drawback of this approach seems to be the large amount of CPU time required for a moderately parameterized model and the possibility of problems in convergence in estimating the bootstrap samples.

5.5.2 An LM test

We see in the previous subsection a possible solution to the model selection problem for nonlinear time series. However, such an approach may not be too convenient to use and could encounter numerical problems. In Li (1993) a simple one degree of freedom test for discriminating among nonlinear models is developed. This new test has some advantage over the bootstrapped Cox test in that it is easy to compute and that it avoids the conceptual problem that faces the bootstrap. More importantly, simulation results suggest that the test statistic has satisfactory power and approximately the correct sizes in large samples. It can also be shown that the test statistics are in some way related to the comparison of residual variances. Hence, the proposed methodology may be regarded as a formalization of the latter procedure.

For simplicity we consider only two possible hypotheses and follow Li (1993) closely. Generalization to the more general case is direct. Denote the time series process by $\{X_t\}$. It is assumed that $\{X_t\}$ is stationary with at least finite second order moments. Let F_t be the σ-field generated by $\{X_t, X_{t-1}, \ldots\}$, and $\{a_{it}\}$, $i = 1, 2$, be Gaussian white noise processes with means zero and variances σ_i^2, $i = 1, 2$. The null and alternative hypotheses are, respectively,

$$H_0 : X_t = f(F_{t-1}; \gamma) + a_{1t} \quad \text{and} \quad H_1 : X_t = g(G_{t-1}; \boldsymbol{\beta}) + a_{2t} \ , \tag{5.43}$$

where the forms of f and g are known and both have continuous second order derivatives with respect to $\boldsymbol{\gamma}$ and $\boldsymbol{\beta}$. Here $\boldsymbol{\gamma}$ and $\boldsymbol{\beta}$ are $p_i \times 1$ vectors of unknown parameters, $i = 1, 2$. To avoid the possibility of unidentifiability it is further assumed that the two families of models $\{f(F_{t-1}, \boldsymbol{\gamma})\}$ and $\{g(F_{t-1}, \boldsymbol{\beta})\}$ are nonoverlapping. That is, $\{f(F_{t-1}, \boldsymbol{\gamma})\} \cap \{g(F_{t-1}, \boldsymbol{\beta})\} = \{\ \}$, the empty set. In the case of bilinear and threshold models this would mean that the possibility of a linear model is excluded. In practice, tests such as those in Saikkonen and Luukkonen (1988) can be employed to see if linear models are adequate. Note that in Vuong (1989) a variance test is suggested in the independent case to check if two families of models can be considered as equivalent. Vuong proposes that if such is the case then no more testing will be needed. Extension of his result to the time series situation is certainly relevant and important but is clearly too involved to be included in the present book. Denote maximum likelihood estimators of $\boldsymbol{\gamma}$ and $\boldsymbol{\beta}$ by $\hat{\boldsymbol{\gamma}}$ and $\hat{\boldsymbol{\beta}}$. Denote the corresponding residuals by $\hat{a}_{it}$, $i = 1, 2$, and let $\tilde{X}_t = g(F_{t-1}; \hat{\boldsymbol{\beta}})$, the prediction of X_t under the alternative model. Consider the model

$$X_t = f(F_{t-1}; \boldsymbol{\gamma}) + \lambda g(F_{t-1}; \hat{\boldsymbol{\beta}}) + a_t \ , \tag{5.44}$$

where $\{a_t\}$ are zero mean Gaussian white noise with variance σ^2. A test of H_0 against the alternative H_1 can be based on testing $H_0' : \lambda = 0$. This test may be interpreted as a test of the adequacy of the null model vs. a possible deviation in the direction of the alternative. Note that McAleer, McKenzie, and Hall (1988) adopted a similar approach for testing a pure moving average model against a pure autoregressive model. The test of H_0' can be based on the Lagrange multiplier approach of §2.5. Let

$S = \sum a_t^2/2\sigma^2$ and $\boldsymbol{\theta} = (\gamma^{\mathrm{T}}, \lambda)^{\mathrm{T}}$. Then the Lagrange multiplier test for $\lambda = 0$ (Li, 1993) is given by

$$T = \left(\frac{\partial S}{\partial \boldsymbol{\theta}}\right)^{\mathrm{T}} \left[E\left(\frac{\partial S}{\partial \boldsymbol{\theta}} \frac{\partial S^{\mathrm{T}}}{\partial \boldsymbol{\theta}}\right)\right]^{-1} \left(\frac{\partial S}{\partial \boldsymbol{\theta}}\right)$$

where the expectation is evaluated under the null hypothesis. Under the null hypothesis T would be asymptotically chi-squared distributed with one degree of freedom. For simplicity, let n be the same as the effective sample size in estimating $\hat{\gamma}$. Since $\partial S/\partial \boldsymbol{\theta} = \sigma^{-2} \sum a_t \partial a_t/\partial \boldsymbol{\theta}$, the statistic T can be rewritten as

$$T = \sigma_1^{-2} \sum \hat{a}_t \left(\frac{\partial a_t}{\partial \gamma^{\mathrm{T}}}, \tilde{X}_t\right) \left[E\left(\sum \frac{\partial a_t}{\partial \boldsymbol{\theta}} \sum \frac{\partial a_t}{\partial \boldsymbol{\theta}^{\mathrm{T}}}\right)\right]^{-1}$$
$$\sum \left(\frac{\partial a_t}{\partial \gamma^{\mathrm{T}}}, \tilde{X}_t\right)^{\mathrm{T}} \hat{a}_t$$

where $\tilde{y}_t = g(F_{t-1}; \hat{\boldsymbol{\beta}})$ and $\partial a_t/\partial \gamma$ is evaluated under H_0, and $\hat{a}_t = \hat{a}_{1t}$. For n large enough we may drop the expectation operator and rewrite T as

$$T' = na^{\mathrm{T}} W^{\mathrm{T}} (WW^{\mathrm{T}})^{-1} Wa / \sum \hat{a}_t^2 \qquad (5.45)$$

where W^{T} is the $n \times (p_1 + 1)$ matrix of regressors formed by stacking $(\partial a_t/\partial \gamma^{\mathrm{T}}, \tilde{X}_t)$ and $a^{\mathrm{T}} = (\hat{a}_1, \ldots, \hat{a}_n)$. The statistic T' will have the same asymptotic distribution as T under H_0. Thus, as in §2.5, the T' statistic can be interpreted as n times the coefficient of determination of the regression of $\hat{a}_{1t}$ on $\partial a_t/\partial \gamma|\hat{\gamma}$ and $\tilde{X}_t$. In other words, the Lagrange multiplier statistic for testing $\lambda = 0$ can be easily obtained from an auxiliary ordinary regression. It is desirable in nonnested testing to interchange the role of the null and the alternative (Cox (1962)). There is, of course, the possibility of having both hypotheses rejected. Although the interpretation problem can be difficult, such a result is still informative in the sense that it may lead us to a better model different from the existing possibilities. Clearly, generalization of the above procedure to the case of more than one alternative is direct. The empirical size and power of T' in discriminating among different nonlinear time series models are considered in Li (1993) using simulation.

The T' statistic is related to the method of comparing residual variances. Consider as in Li (1993) two auxiliary regressions

$$\hat{a}_{1t} = \tau \tilde{X}_t + \epsilon_t \qquad (5.46)$$

and

$$\tilde{X}_t = \frac{\partial f(F_{t-1}; \gamma)}{\partial \gamma} \cdot \boldsymbol{K} + V_t \ , \qquad (5.47)$$

where ϵ_t, V_t are independent zero mean normal random variates; τ and $\boldsymbol{K}$ are the respective regression parameters. For simplicity, let $\sigma_1^2 = 1$. Then under H_0 the score vector $\partial S/\partial \boldsymbol{\theta} = -(\mathbf{0}, \sum \hat{a}_{1t}\tilde{X}_t)$ and the observed Fisher information matrix,

$$I = \begin{bmatrix} \sum \frac{\partial f_t}{\partial \gamma}\frac{\partial f_t}{\partial \gamma^{\mathrm{T}}} & \sum \tilde{X}_t \frac{\partial f_t}{\partial \gamma} \\ \sum \tilde{X}_t \frac{\partial f_t}{\partial \gamma^{\mathrm{T}}} & \sum \tilde{X}_t^2 \end{bmatrix},$$

where $f_t = f(F_{t-1};\gamma)$. Hence, the statistic T' can be written as

$$\begin{aligned} T' &= (\sum \hat{a}_{1t}\tilde{X}_t)^2 \left[\sum \tilde{X}_t^2 - \left(\sum \tilde{X}_t \frac{\partial f_t}{\partial \gamma^{\mathrm{T}}}\right)\left(\sum \frac{\partial f_t}{\partial \gamma}\cdot\frac{\partial f_t}{\partial \gamma^{\mathrm{T}}}\right)^{-1} \right. \\ &\quad \left.\left(\sum \tilde{X}_t \frac{\partial f_t}{\partial \gamma}\right)\right]^{-1} \\ &= \left[\frac{\sum \hat{a}_{1t}\tilde{X}_t}{\sum \tilde{X}_t^2}\right]^2 \cdot \frac{\sum \tilde{X}_t^2}{[1-r^2]}, \end{aligned}$$

where

$$r^2 = \frac{\left(\sum \tilde{X}_t \frac{\partial f_t}{\partial \gamma^{\mathrm{T}}}\right)\left(\sum \frac{\partial f_t}{\partial \gamma}\cdot\frac{\partial f_t}{\partial \gamma^{\mathrm{T}}}\right)^{-1}\left(\sum \tilde{X}_t \frac{\partial f_t}{\partial \gamma}\right)}{\sum \tilde{X}_t^2}.$$

The quantity r^2 is the coefficient of determination for the auxiliary regression (5.47). Note that $\sum \hat{a}_{1t}\tilde{X}_t / \sum \tilde{X}_t^2 = \hat{\tau}$, the least squares estimate of τ in (5.46). Hence, using standard regression results

$$\begin{aligned} T' &= \frac{\sum \hat{\tau}^2 \tilde{X}_t^2}{1-r^2} \\ &= \frac{\sum \hat{a}_{1t}^2 - \sum \hat{\epsilon}_t^2}{1-r^2}. \end{aligned} \tag{5.48}$$

We observe from (5.48) that if H_0 is the true model then $\sum \hat{a}_{1t}^2$ should be small and $\sum \hat{\epsilon}_t^2$ should be closed to $\sum \hat{a}_{1t}^2$. However, if H_1 is true then $\sum \hat{a}_{1t}^2$ should be large while $\sum \hat{\epsilon}_t^2$ should be small. A similar result holds when we interchange the hypotheses. Thus the testing procedure can be interpreted as a way to compare residual variances after adjusting them by auxiliary regressions (5.46) and (5.47). One advantage of the approach is, clearly, that the statistic T' has a known asymptotic distribution under the null hypothesis and therefore we can have meaningful discussions on sizes and power at least asymptotically. The parameter r^2 can be interpreted as a measure of the similarity between $g(F_{t-1};\boldsymbol{\beta})$ and

$f(F_{t-1};\gamma)$ since, in the special case, where $g(F_{t-1};\beta) = \beta g(F_{t-1})$ and $f(F_{t-1};\gamma) = \gamma f(F_{t-1})$, then $r^2 = 1$ if $cf = g$ for some constant c. Note also that since $0 < r^2 < 1$, the test statistic can be much larger than its numerator and hence the procedure can be more sensitive in detecting significant differences of the models than the method of comparing residual variances.

Example 5.4 *The Wölf sunspot numbers (Li, 1993). Reproduced with the permission of Academic Sinica, Taipei*

As a real example we considered again the annual Wölf sunspot numbers (1700–1921). Since in example 5.3 $\hat{\phi}_4$ *and* $\hat{\phi}'_{12}$ *are actually not significant we consider here the* SETAR *(2; 3, 11) model in Tong (1990, p. 425) and the subset bilinear model of Gabr and Subba Rao (1981). These nonlinear models were refitted by considering the first eleven observations as fixed and two* T'_1 *statistics* $\tilde{T}_1$ *and* $\tilde{T}_2$ *were computed. The* $\tilde{T}_1$ *statistic had the threshold model as the null and the subset bilinear model as the alternative and the* $\tilde{T}_2$ *statistic had the hypotheses the other way around. The refitted models and the* $\tilde{T}_i$ *statistics are as follows. For the threshold model we had*

$$X_t = \begin{cases} 10.7678 + 1.7344X_{t-1} - 1.2957X_{t-2} + 0.4740X_{t-3} + \epsilon_t \ , \\ \hfill \text{if } X_{t-3} \le 36.6 \ , \\ 7.5791 + 0.7332X_{t-1} - 0.0403X_{t-2} - 0.1971X_{t-3} \\ \quad + 0.1597X_{t-4} - 0.2204X_{t-5} + 0.0220X_{t-6} + 0.1491X_{t-7} \\ \quad - 0.2403X_{t-8} + 0.3121X_{t-9} - 0.3691X_{t-10} + 0.3881X_{t-11} \\ \quad + \epsilon_t \ , \hfill \text{if } X_{t-3} > 36.6 \end{cases}$$

and $\tilde{T}_1 = 51.84$. *Note that here the residuals for both branches of the model were taken to have the same variance. For the subset bilinear model we had*

$$\begin{aligned} X_t \;\; = \;\; & 6.8922 + 1.5012X_{t-1} - 0.7671X_{t-2} + 0.1152X_{t-9} \\ & - 0.0146X_{t-2}e_{t-1} + 0.0063X_{t-8}e_{t-1} \\ & - 0.0072X_{t-1}e_{t-3} + 0.0068X_{t-4}e_{t-3} \\ & + 0.0036X_{t-1}e_{t-6} + 0.0043X_{t-2}e_{t-4} \\ & + 0.0018X_{t-3}e_{t-2} + e_t \end{aligned}$$

and $\tilde{T}_2 = 0.0268$. *Hence, the* $\tilde{T}_1$ *statistic rejected the threshold null while the* $\tilde{T}_2$ *statistic accepted the bilinear null. Thus the approach here favored the bilinear model over the threshold model for the time period considered. The residual variances for the bilinear and threshold models were respectively 124.92 and 149.71. Note that the value of 0.0268, although small, was still greater than the lower 10% critical value of a chi-square distribution with one degree of freedom. This example also reflects the dependence of the test on the residual variance. Clearly predictive power is not the only criterion for choosing a nonlinear model.*

CHAPTER 6

Conditional heteroscedasticity models

6.1 The autoregressive conditional heteroscedastic model

Just about the time that the nonlinear time series models were being developed time series analysis in econometrics took to another path of development. This development occurred because of the need to model data in economics and in particular, in finance where heteroscedasticity is the norm. Hence, the autoregressive moving average (ARMA) model with Gaussian noise and constant variance is inadequate in describing such data. Consider the classical regression model

$$y_t = \boldsymbol{X}_t^{\mathrm{T}} \boldsymbol{\beta} + \epsilon_t \ , \tag{6.1}$$

where $\boldsymbol{\beta}$ is a $p \times 1$ vector of regression parameters, $\{\epsilon_t\}$ is an independent noise sequence, and $\boldsymbol{X}_t$ is a $p \times 1$ vector of explanatory variables. The classical solution to the heteroscedasticity problem is to assume that the variance of ϵ_t is given by $\sigma^2 Z_{t-1}$ where Z_{t-1} is an exogenous variable. As argued by Engle (1982) this solution is unsatisfactory in the time series context as it fails to recognize that the variance, like the mean, can also evolve over time. Let the time series be denoted $\{y_t\}$. Denote by F_{t-1} all the information available up to time $t-1$. In many situations we consider only the time series y_t itself and hence $F_{t-1} = \{y_{t-1}, \cdots\}$. Engle (1982) proposed that the conditional variance of ϵ_t can be modeled as

$$\epsilon_t = \sqrt{h_t}\, a_t \tag{6.2}$$

where

$$h_t = h(y_{t-1}, \ldots, y_{t-q}, \boldsymbol{\alpha}) \ . \tag{6.3}$$

Here $h(\ )$ is a non-negative function of past y_t's, $\boldsymbol{\alpha}$ a $q \times 1$ vector of parameters, and a_t are independent identically distributed white noise with mean 0 and variance 1. In many applications, in particular in financial time series,

$$y_t = \epsilon_t = \sqrt{h_t}\, a_t \ . \tag{6.4}$$

This will be assumed from now on unless otherwise stated. In this case $F_{t-1} = \{y_{t_1}, \cdots\} = \{\epsilon_{t-1}, \cdots\}$. The unconditional mean of ϵ_t is, from (6.2),

$$\begin{aligned} E(\epsilon_t) &= E(\sqrt{h_t}\, a_t) \\ &= E(\sqrt{h_t})E(a_t) \\ &= 0 \end{aligned}$$

because a_t is independent of $\sqrt{h_t}$ and $E(a_t) = 0$. Furthermore, the conditional variance of ϵ_t given past ϵ_t's is just

$$\begin{aligned} E(\epsilon_t^2 \mid F_{t-1}) &= E(h_t\, a_t^2 \mid F_{t-1}) \\ &= E(h_t \mid F_{t-1}) \cdot E(a_t^2) \\ &= h_t\ . \end{aligned}$$

There are many possible ways to define $h(\cdot)$ but a simple expression is the specification

$$h_t = \alpha_0 + \alpha_1 \epsilon_{t-1}^2\ , \tag{6.5}$$

with $\alpha_0 > 0$, $\alpha_1 \geq 0$.

In this specification, the conditional variance h_t is dependent on ϵ_{t-1}^2, where ϵ_{t-1} is the previous shock or noise to the time series. Hence a large previous shock ϵ_{t-1} will lead to a larger conditional variance for $\epsilon_t(y_t)$. This specification seems to match well with the empirical observations in economic and financial time series where a large ϵ_{t-1}, caused by news arrivals to the market, could generate successive large fluctations in subsequent periods. Engle (1982) called models (6.2) and (6.5) a first order autoregressive conditional heteroscedastic (ARCH(1)) process. A higher order (ARCH(q)) process can be defined by including more past ϵ_t's, that is,

$$h_t = \alpha_0 + \alpha_1 \epsilon_{t-1}^2 + \cdots + \epsilon_{t-q}^2 \tag{6.6}$$

where $\alpha_0 > 0$, $\alpha_i \geq 0$, $i = 1\ldots, q$. Note that for the ARCH(1) process (6.5), we have, assuming second order stationarity for ϵ_t, that

$$\begin{aligned} \mathrm{var}(\epsilon_t) = E(\epsilon_t^2) &= E(h_t) \\ &= E(\alpha_0 + \alpha_1 \epsilon_{t-1}^2) \\ &= \alpha_0 + \alpha_1 E(\epsilon_{t-1}^2) = \alpha_0 + \alpha_1 E(\epsilon_t^2)\ . \end{aligned} \tag{6.7}$$

Consequently,

$$\mathrm{var}(\epsilon_t) = E(\epsilon_t^2) = \frac{\alpha_0}{1 - \alpha_1}\ . \tag{6.8}$$

The above result also suggests that the condition for second order stationarity is $\alpha_1 < 1$. In the financial market large falls and rises in an asset's price P_t are often observed. As a result the empirical distribution of the return series $R_t = \ln P_t - \ln P_{t-1}$ often has tails fatter than those of the normal distribution. It is therefore of interest to see if the ARCH(q) models can mimic this feature.

Suppose now that a_t is standard normal and for simplicity let $q = 1$. Then the fourth order moment of ϵ_t is given by

$$\begin{aligned} E(\epsilon_t^4) &= E(h_t^2 a_t^4) \\ &= 3\,E(h_t^2) \end{aligned} \tag{6.9}$$

because $E(a_t^4) = 3$.

Now

$$\begin{aligned} E(h_t^2) &= E(\alpha_0 + \alpha_1 \epsilon_{t-1}^2)^2 \\ &= E(\alpha_0^2 + 2\alpha_0\alpha_1\epsilon_{t-1}^2 + \alpha_1^2\epsilon_{t-1}^4) \\ &= \alpha_0^2 + 2\alpha_0\alpha_1 E(h_{t-1}) + \alpha_1^2 E(\epsilon_{t-1}^4) \\ &= \alpha_0^2 + 2\alpha_0\alpha_1 \frac{\alpha_0}{1-\alpha_1} + \alpha_1^2 E(\epsilon_t^4)\ . \end{aligned} \tag{6.10}$$

The last line requires stationarity to the fourth order. Substituting (6.10) into (6.9) gives

$$E(\epsilon_t^4) = 3\alpha_0^2 + 6\alpha_0^2\alpha_1/(1-\alpha_1) + 3\alpha_1^2 E(\epsilon_t^4)\ .$$

Thus

$$\begin{aligned} E(\epsilon_t^4) &= \frac{3\alpha_0^2}{1-3\alpha_1^2}\left(1 + \frac{2\alpha_1}{1-\alpha_1}\right) \\ &= \frac{3\alpha_0^2}{1-3\alpha_1^2}\frac{1+\alpha_1}{(1-\alpha_1)} \\ &= \frac{3\alpha_0^2}{1-3\alpha_1^2}\frac{(1-\alpha_1^2)}{(1-\alpha_1)^2}\ . \end{aligned}$$

Hence, the fourth order moments of ϵ_t exists if $1-3\alpha_1^2 > 0$ or equivalently if $\alpha_1^2 < 1/3$. Furthermore, if we consider the kurtosis K_4 of ϵ_t we have

$$\begin{aligned} K_4 &= \frac{E(\epsilon_t^4)}{E(\epsilon_t^2)^2} - 3 \\ &= \frac{3\alpha_0^2}{1-3\alpha_1^2}\frac{(1-\alpha_1^2)}{(1-\alpha_1)^2}\frac{(1-\alpha_1)^2}{\alpha_0^2} - 3 \\ &= \frac{3(1-\alpha_1^2)}{1-3\alpha_1^2} - 3 > 0\ , \end{aligned} \tag{6.11}$$

because $1 - 3\alpha_1^2 < 1 - \alpha_1^2$. The above result implies that the distribution of ϵ_t has tails fatter than those of the normal distribution, an empirical fact with many financial return series, where return is defined as the first order difference of the logarithmic transformed series. Note that for the general ARCH(p) process the stationary variance can be shown to be $\alpha_0/(1-\alpha_1\cdots-\alpha_p)$.

Estimation of the ARCH(q) process can be achieved via the method of maximum likelihood by assuming that ϵ_t is conditionally normally distributed. That is,

$$\epsilon_t|F_{t-1} \sim N(0, h_t)$$

where

$$h_t = h(\epsilon_{t-1}, \ldots, \epsilon_{t-q}) ,$$

which is equal to $\alpha_0 + \alpha_1\epsilon_{t-1}^2 + \cdots + \epsilon_{t-q}^2$ in the ARCH(q) case. The log-likelihood function at time t, l_t is given by

$$l_t = -\frac{1}{2}\log h_t - \frac{1}{2}y_t^2/h_t \qquad (6.12)$$

and the log-likelihood function l for a realization of length n conditional on the first q observations is just

$$l = \sum_{t=q+1}^{n} l_t .$$

In many econometric applications a t distribution is assumed for a_t in (6.2) resulting in even fatter tails for the process y_t. It may be shown that if ϵ_t follows the more general models (6.1) and (6.3) then the information matrix for the parameters $\boldsymbol{\beta}$ and $\boldsymbol{\alpha}$ are block diagonal under some general conditions (Engle, 1982). This implies that during the estimation process we can have two separate sets of estimating equations for $\boldsymbol{\beta}$ and $\boldsymbol{\alpha}$, respectively, and that the estimates $\hat{\boldsymbol{\beta}}$ and $\hat{\boldsymbol{\alpha}}$ are asymptotically independent of each other. Many authors recommended the use of the so-called BHHH algorithm (Berndt, Hall, Hall, and Hausman, 1974) in finding the log-likelihood estimates. This algorithm only requires the first order derivatives of l_t with respect to the parameters. However, this approach may have numerical problems under certain situations. In Mak, Wong, and Li (1997) an iteratively weighted least squares scheme is suggested which provides better convergence properties than the BHHH algorithm.

The ARCH models were first applied to study the variance of UK inflation by Engle (1982) and to the US inflation by Engle (1983). A huge literature now exists for the ARCH models. Bollerslev, Chou, and Kroner (1992) and Bollerslev, Engle, and Nelson (1994) are two earlier reviews while Li, Ling, and McAleer (2002) gave a more recent update.

Bollerslev (1986) extended the ARCH(q) process by including lagged values of h_t. The generalized autoregressive conditional heteroscedastic

(GARCH) model of order (p, q) is defined by

$$\begin{aligned} \epsilon_t \mid F_{t-1} &\sim N(0, h_t) \\ h_t &= E(\epsilon_t^2 \mid F_{t-1}) = \alpha_0 + \sum_{i=1}^{q} \alpha_i \epsilon_{t-i}^2 + \sum_{i=1}^{p} \beta_i h_{t-i} \ , \end{aligned} \tag{6.13}$$

where $\alpha_i > 0$, $\alpha_i \geq 0$, $i = 1, \ldots, q$, and $\beta_i \geq 0$, $i = 1, \ldots, p$. Clearly for $p = 0$, (6.13) becomes the usual ARCH(q) process. Note that the inequality constraints on α_i and β_i can be weakened (Nelson and Cao, 1991).

Let

$$A(B) = \sum_{i=1}^{q} \alpha_i B^i$$

and

$$C(B) = \sum_{i=1}^{p} \beta_i B^i \ ,$$

where B denotes the backward shift operator, then the condition for covariance stationarity for (6.13) is that

$$A(1) + C(1) < 1$$

with stationary variance given by $\alpha_0(1 - A(1) - B(1))^{-1}$. By subtracting h_t from ϵ_t^2 we have

$$\epsilon_t^2 - h_t = \epsilon_t^2 - \alpha_0 - \sum_{i=1}^{q} \alpha_i \epsilon_{t-i}^2 - \sum_{i=1}^{p} \beta_i h_{t-i} \ .$$

Adding and subtracting the terms $\beta_i \epsilon_{t-i}^2$, $i = 1, \ldots, q$ on the right-hand side gives

$$\begin{aligned} \epsilon_t^2 - h_t &= \epsilon_t^2 - \alpha_0 - \sum_{i=1}^{q} \alpha_i \epsilon_{t-i}^2 - \sum_{i=1}^{p} \beta_i \epsilon_{t-i}^2 \\ &\quad - \sum_{i=1}^{p} \beta_i (h_{t-i} - \epsilon_{t-i}^2) \ . \end{aligned} \tag{6.14}$$

Write $V_t = \epsilon_t^2 - h_t$ and by setting α_i to 0, $p \geq i > q$, if $p > q$ (or if $q > p$, set β_i to 0, $q \geq i > p$) (6.14) can be written as

$$\epsilon_t^2 = V_t + \sum_{i=1}^{\max(p,q)} (\alpha_i + \beta_i) \epsilon_{t-i}^2 + \sum_{i=1}^{p} \beta_i V_{t-i} \ . \tag{6.15}$$

Since V_t can be regarded as white noise (6.15) suggests that ϵ_t^2 satisfies an ARMA(P, Q) representation with autoregressive order $P = \max(p, q)$ and moving average order $Q = p$. The most successful GARCH model appears to be the GARCH(1, 1) model. Again the GARCH(1, 1) model has an excess kurtosis greater than 0 and its distribution is therefore also heavy-tailed as in the ARCH(1) case. Bollerslev (1986) applied the GARCH(1, 1) model to the rate of growth of the US implicit GNP deflator.

In many applications an AR or ARMA component is often considered for the conditional mean of the series y_t. In the former case we have

$$y_t = \phi_0 + \phi_1 y_{t-1} + \cdots + \phi_p y_{t-p} + \epsilon_t$$

where ϵ_t is an ARCH(q) or GARCH(p, q) process. In terms of (6.1) this amounts to having $\boldsymbol{X}_t = (1, y_{t-1} \cdots y_{t-p})^{\mathrm{T}}$ and $\boldsymbol{\beta} = (\phi_1, \ldots, \phi_p)^{\mathrm{T}}$. Extension to ARMA-ARCH is direct. Asymptotic theory and estimation for ARMA-ARCH models are given by Weiss (1986). Weiss (1986) also studied the case where the log-likelihood l_t used in (6.12) is not the true log-likelihood for ϵ_t and therefore the estimates obtained are only quasi-likelihood estimates.

6.2 Checks for the presence of ARCH

In this section we consider tests for the possible presence of ARCH.

(i) A Lagrange multiplier (LM) test with a portmanteau equivalent

Engle (1982) originally derived an LM test for the presence of ARCH. Let $\hat{\epsilon}_t$ be the residuals from a least squares fit of the model

$$y_t = \phi_0 + \phi_1 y_{t-1} + \cdots + \phi_p y_{t-p} + \epsilon_t \ .$$

Let $\boldsymbol{z}_t = (1, \hat{\epsilon}^2_{t-1}, \ldots, \hat{\epsilon}^2_{t-q})$ and let $h_t = h(z_t \, \boldsymbol{\alpha})$, where $\boldsymbol{\alpha}$ is a $(q+1)$ vector of parameters. Under the null of no autoregressive conditional heteroscedasticity h_t is a constant equal to h_0.

Assuming a normal ϵ_t Engle's LM test in the sense of §(2.5) is given by

$$\mathrm{LM} = \frac{1}{2} \boldsymbol{f}_0^{\mathrm{T}} \boldsymbol{z} (\boldsymbol{z}^{\mathrm{T}} \boldsymbol{z})^{-1} \boldsymbol{z}^{\mathrm{T}} \boldsymbol{f}_0 \qquad (6.16)$$

where $\boldsymbol{z}^{\mathrm{T}} = (\boldsymbol{z}^{\mathrm{T}}_{p+1}, \ldots, \boldsymbol{z}^{\mathrm{T}}_n)$ and

$$\boldsymbol{f}_0 = [(\hat{\epsilon}^2_{p+1}/h_0 - 1), \ldots, (\hat{\epsilon}^2_n/h_0 - 1)]^{\mathrm{T}} \ .$$

LM is asymptotically χ^2_q distributed under the null hypothesis of no ARCH. An asymptotically equivalent form of LM can be obtained by regressing $\hat{\epsilon}^2_t$ on $(1, \hat{\epsilon}^2_{t-1}, \ldots, \hat{\epsilon}^2_{t-q})^{\mathrm{T}}$ and then the test is given by $n \cdot R^2$, the coefficient of determination of this regression.

Luukkonen, Saikkonen, and Teräsvirta (1988) pointed out that the LM test is, in fact, asymptotically equivalent to the McLeod-Li portmanteau test (5.9) based on the autocorrelations of squared residuals. In the case of $q = 1$ this can be easily seen as follows. Without loss of generality let $y_t = \epsilon_t$ and hence $\boldsymbol{z}_t = (1, \epsilon_{t-1}^2)$. Hence,

$$\begin{aligned} \boldsymbol{f}_0^{\mathrm{T}} \boldsymbol{z} &= \left(\frac{\epsilon_1^2}{h_0} - 1, \cdots, \frac{\epsilon_n^2}{h_0} - 1 \right) \cdot \begin{pmatrix} \boldsymbol{z}_1 \\ \vdots \\ \boldsymbol{z}_n \end{pmatrix} \\ &= \sum_{i=1}^{n} \left(\frac{\epsilon_i^2}{h_0} - 1 \right) \cdot \boldsymbol{z}_i \\ &= \left(\sum_{i=1}^{n} \left(\frac{\epsilon_i^2}{h_0} - 1 \right), \sum_{i=1}^{n} \left(\frac{\epsilon_i^2}{h_0} - 1 \right) \epsilon_{i-1}^2 \right) . \end{aligned} \tag{6.17}$$

Since $E(\epsilon_i^2/h_0) = 1$ we see that the first term when divided by n converges to zero and that the second term when divided by n and h_0 is asymptotically equivalent to $1/n\Sigma(\epsilon_i^2/h_0 - 1)(\epsilon_{i-1}^2/h_0 - 1)$, the lag one autocovariance of ϵ_i^2/h_0, which up to a scaling factor is an alternative expression of (5.7). Further algebra shows that the LM test is asymptotically the McLeod-Li portmanteau test. Therefore, the test statistic (5.9) is not just a pure significance test but an LM test for the presence of ARCH. Advantages of the test (5.7) is clearly its simplicity and that it can be easily programmed.

The above LM test is for the null hypothesis of no ARCH against the alternative of ARCH(q). For testing the null of no ARCH against the alternative of a GARCH(p, q), Lee (1991) showed that the LM test is in fact equivalent to that of testing the same null hypothesis against an ARCH(q) process as the alternative.

(ii) Lee and King's test

In the previous section the LM test for the null of no ARCH against the alternative of an ARCH process ignores the inequality constraints for α_i and β_i. It is natural to ask whether a test with these constraints taken into consideration would have a better performance in terms of size and power. For simplicity let $y_t = \epsilon_t = \sqrt{h_t}\, a_t$. We adopt the same notation of §2.5 where $\boldsymbol{\theta} = (\boldsymbol{\theta}_1^{\mathrm{T}}, \boldsymbol{\theta}_2^{\mathrm{T}})^{\mathrm{T}}$ and $H_0 : \boldsymbol{\theta}_2 = 0$ but the alternative H_A' is now that at least one of the elements of $\boldsymbol{\theta}_2$ is greater than zero. King and Wu (1990) observed that locally most mean powerful (LMMP) tests for these pair of hypotheses has the form

$$S = \sum_{i=1}^{r} \left. \frac{\partial \ln f(x|\boldsymbol{\theta})}{\partial \theta_{2i}} \right|_{\boldsymbol{\theta} = (\boldsymbol{\theta}_1^{\mathrm{T}}, \boldsymbol{0}^{\mathrm{T}})^{\mathrm{T}}} > C \, . \tag{6.18}$$

This test maximizes the mean slope of the power hypersurface in the neighorhood of the null hypothesis H_0. In practice, $\boldsymbol{\theta}_1$ will be replaced by its maximum likelihood estimate $\hat{\boldsymbol{\theta}}_1$ under H_0. A one-sided LM test could be based on the statistic

$$T = \hat{S}/(\imath^{\mathrm{T}}(\hat{I}^{22})^{-1}\imath)^{1/2} \tag{6.19}$$

where I^{22} denotes the lower $r \times r$ block of the inverse of the Fisher information matrix, and $\hat{I}^{22}$ is the value of I^{22} evaluated at $\hat{\boldsymbol{\theta}} = (\hat{\boldsymbol{\theta}}_1^{\mathrm{T}}, \mathbf{0}^{\mathrm{T}})^{\mathrm{T}}$. Similarly $\hat{S}$ is the value of S evaluated at $\hat{\boldsymbol{\theta}}$; $\imath$ is an $r \times 1$ vector of ones. For testing H_0 of no ARCH against the alternative H_A' that at least one of $\alpha_i > 0$, $i = 1, \ldots, q$ the LMMP test has the form (Lee and King, 1993),

$$S_{\mathrm{ARCH}} = \frac{(n-q)\sum\limits_{t=q+1}^{n}(y_t^2/h_0 - 1)\sum\limits_{i=1}^{q} y_{t-i}^2}{\left\{2\sum\limits_{t=q+1}^{n}\Big(\sum\limits_{i=1}^{q} y_{t-i}^2\Big)^2 - 2\Big(\sum\limits_{t=q+1}^{n}\sum\limits_{i=1}^{q} y_{t-i}^2\Big)^2/(n-q)\right\}^{1/2}}. \tag{6.20}$$

A robustified version of (6.20) based on the result of Koenker (1981) is also suggested in Lee and King (1993). Under H_0, S_{ARCH} is asymptotically $N(0,1)$ distributed so that the one-sided test can be easily applied. Simulations in Lee and King (1993) showed that either (6.20) or its robustified version have power that dominates the corresponding LM tests (6.16) and their asymptotic version using $n \cdot R^2$. Assuming that a result of Self and Liang (1987) can be applied to dependent observations, Demos and Sentana (1998) proposed a one-sided LM test which is also more powerful than the two-sided LM test. In the ARCH(1) case, the $n \cdot R^2$ form of the test is obtained as in the two-sided case but H_0 is only rejected when the least squares slope coefficient of regressing $\hat{a}_t^2$ on $\hat{a}_{t-1}^2$ is positive and $nR^2 > 2.706$. Hong (1997) considered a one-sided test based on the a weighted sum of sample autocorrelations of squared regression residuals which has Lee and King's test as a special case.

(iii) Hong's test

Under the null of no ARCH effect, $h_t = h_0$ a constant. Hence ϵ_t^2/h_0 has mean one and is uncorrelated over time. Let $u_t = \epsilon_t^2/h_0 - 1$. Then u_t is a zero mean white noise process. The normalized spectral density $f(w)$ of u_t is $f(w) = f_0(w) = 1/2\pi$ for all frequencies $\omega \in [-\pi, \pi]$. When ARCH is present $f(w) \neq 1/2\pi$ in general. Hong (1996b) proposed a test based on the normalized spectal density of u_t and the L^2 norm. It has the form

$$L_2(\hat{f}; f_0) = \left[2\pi\int_{-\pi}^{\pi}\left[\hat{f}(w) - f_0(w)\right]^2 dw\right]^{1/2}. \tag{6.21}$$

The sample spectral density can be estimated by

$$\hat{f}(w) = (2\pi)^{-1} \sum_{j=1-n}^{n-1} k(j/b)\hat{\rho}(j)\cos(jw)$$

where $w \in [-\pi, \pi]$; $\hat{\rho}(j)$ is the lag j sample autocorrelation of u_t; $b = b(n)$ is a bandwidth such that $q \to \infty$, $q/n \to 0$ as n goes to infinity; $k : R \to [-1, 1]$ is a symmetric kernel function, continuous at 0 with $k(0) = 1$; and $\int_{-\infty}^{-\infty} k^2(z)dz < \infty$. Hong and Shehadeh (1999) defined the test statistic

$$Q(b) = \frac{\left(\frac{n}{2}L_2^2(\hat{f}; f_0) - C_n(k)\right)}{2(D_n(k))^{1/2}} \tag{6.22}$$

where

$$C_n(k) = \sum_{j=1}^{n-1}(1 - j/n)k^2(j/b)$$

and

$$D_n(k) = \sum_{j=1}^{n-2}(1 - j/n)(1 - (j+1)/n)k^4(j/b)\ .$$

The test statistic $Q(b)$ can be written

$$Q(b) = \frac{\left(n\sum_{j=1}^{n-1} k^2(j/b)\hat{\rho}(j)^2 - C_n(k)\right)}{(2D_n(k))^{1/2}}\ .$$

Replacing $C_n(k)$ by $bC(k)$ where $C(k) = \int_0^{\infty} k^2(z)d(z)$ we have asymptotically an equivalent test

$$Q^*(b) = \frac{\left(n\sum_{j=1}^{n-1} k^2(j/b)\hat{\rho}(j)^2 - bC(k)\right)}{(2bD(k))^{1/2}} \tag{6.23}$$

where $D(k) = \int_0^{\infty} k^4(z)d(z)$.

Hong and Shehadeh (1999) proposed a cross-validation procedure for the choice of the bandwidth b. He also demonstrated the relationship of (6.23) with various tests for ARCH by using different kernels $k(\cdot)$. For example, if k is the truncated kernel $k(z) = 1$ for $|z| \leq 1$ and 0 for $|z| > 1$, $Q(b)$ becomes

$$Q_{\text{trun}}(b) = \frac{Q_{aa} - b}{(2b)^{1/2}} \tag{6.24}$$

where

$$Q_{aa} = n \cdot \sum_{j=1}^{b} \hat{\rho}(j)^2 .$$

Thus (6.24) is asymptotically equivalent to the Q_{aa}^* statistic (5.9). The simulation in Hong and Shehadeh (1999) suggested that for the Daniell kernel with or without cross-validation the $Q(b)$ statistic in general performs reasonably well when compared with other statistics. The price to pay is, of course, the heavier computational burden in computing the statistic and in choosing the bandwidth.

Hong and Shehadeh (1999) also proposed an alternative Ω_b to (6.22) based on the supremum norm,

$$\begin{aligned}\Omega_b &= \left(\frac{n}{2}\right)^{1/2} \sup_{w\in[0,\pi)} \left|\hat{f}(w) - f_0(w)\right| \\ &= n^{1/2} \sup_{w\in[0,\pi]} \left|\sum_{j=1}^{n-1} k(j/b)\hat{\rho}(j)\sqrt{2}\cos(jw)\right| . \qquad (6.25)\end{aligned}$$

(iv) A rank portmanteau statistic

With the possible presence of outliers rank autocorrelations are attractive non-parametric alternatives to standard autocorrelation coefficients. Though many definitions have appeared in the literature, the most natural definition for the rank autocorrelation at lag k for a time series $\{y_1, \ldots, y_n\}$ seems to be

$$\tilde{r}_k = \frac{\sum\limits_{t=k+1}^{n} (R_t - \bar{R})(R_{t-k} - \bar{R})}{\sum\limits_{t=1}^{n} (R_t - \bar{R})^2}, \qquad 1 \leq k \leq n-1 \qquad (6.26)$$

where R_t is the rank of observation y_t, with

$$\begin{aligned}\bar{R} = \sum_{t=1}^{n} R_t/n &= n(n+1)/2 \\ \sum_{t=1}^{n} (R_t - \bar{R})^2 &= n(n^2-1)/12\end{aligned}$$

Dufour and Roy (1985, 1986) showed that the distribution of the rank autocorrelations is the same whenever $y_1, \ldots, y_n$, are continuous exchangeable random variables. The reason is that all rank permutations in this situation are equally probable. Moran (1948) first showed that

$E(\tilde{r}_k) = -(n-k)/n(n-1)$. Dufour and Roy (1986) further showed that

$$\text{var}(\tilde{r}_k) = \frac{5n^4 - (5k+9)n^3 + 9(k-2)n^2 + 2k(5k+8)n + 16k^2}{5(n-1)^2n^2(n+1)}, \quad 1 \le k \le n-1 \tag{6.27}$$

Finally, letting $\mu_k = E(\tilde{r}_k)$ and $\tilde{\sigma}_k^2 = \text{var}(\tilde{r}_k)$, Dufour and Roy (1986) showed that the statistic

$$Q_R = \sum_{k=1}^{M} \frac{(\tilde{r}_k - \mu_k)^2}{\tilde{\sigma}_k^2} \tag{6.28}$$

follows a χ_M^2 distribution asymptotically.

It is easy to see that squared residuals correspond to continuous exchangeable random variables asymptotically. Thus, in (6.26), if R_t is the rank of the squared residual, i.e., $R_t = \text{rank}(\hat{a}_t^2)$, Q_R of (6.28) is a portmanteau statistic of the ranks of squared residuals. Q_R is the rank version of the McLeod-Li statistic and Q_R follows a χ_M^2 distribution asymptotically (Wong and Li, 1995).

Some simulation experiments were considered in Wong and Li (1995) for the AR(1) model

$$y_t = \phi y_{t-1} + a_t \tag{6.29}$$

where $t = 1, \ldots, n$, $n = 50, 200$, and $\phi = 0, \pm 0.3, \pm 0.6, \pm 0.9$. The a_t terms are independent $N(0, \sigma_a^2)$ random variables. Here $\sigma_a^2 = 1$. Each of the models were simulated 1000 times using IMSL subroutines. The empirical p values of Q_R at the asymptotic upper 5% level are shown in Table 6.1. Here, the degrees of freedom are $M = 1, 4, 7$, and 10. Note that the critical values of χ_1^2, χ_4^2, χ_7^2, and χ_{10}^2 at 5% are 3.841, 9.488, 14.067, and 18.307, respectively.

To investigate the robustness of Q_R, the simulations were repeated in Wong and Li (1995) with three randomly assigned outliers to the generated series. Each outlier is equal to $\mu + 3\sigma_a$, i.e., 3. As a comparison, similar experiments were performed with Q_{aa}^* (5.9), and the results are shown in Table 6.2.

From Table 6.1, the overall empirical significance level of Q_R is close to 5% when there is no outlier. Similar conclusions can be drawn from Table 6.2 where there are three outliers. The 5% critical values of Q_R appear to be only slightly affected. These results indicate that the finite sample distribution is robustly approximated by the asymptotic distribution for the sample sizes and degrees of freedom under consideration. However, from Table 6.2, it is observed that the empirical size of Q_{aa}^* changes quite dramatically in the presence of outliers. The performance

Table 6.1 *Empirical p values of Q_R at 5% level (Wong and Li, 1995). Reproduced with the permission of Taylor & Francis Ltd.*

No. of observations (n)	ϕ	No. of outliers	Degrees of freedom (M) 1	4	7	10
50	−0.9	0	0.047	0.051	0.064	0.076
	−0.6		0.049	0.049	0.058	0.072
	−0.3		0.052	0.054	0.065	0.067
	0		0.054	0.052	0.060	0.071
	0.3		0.053	0.055	0.065	0.076
	0.6		0.053	0.055	0.057	0.071
	0.9		0.047	0.054	0.060	0.069
200	−0.9	0	0.054	0.046	0.049	0.055
	−0.6		0.053	0.047	0.048	0.054
	−0.3		0.056	0.045	0.050	0.054
	0		0.054	0.044	0.050	0.053
	0.3		0.053	0.042	0.052	0.053
	0.6		0.054	0.044	0.050	0.055
	0.9		0.048	0.045	0.050	0.055

of Q^*_{aa} and Q_R under the ARCH model of order one was also considered by Wong and Li (1995). The model is

$$y_t = (\alpha_0 + \alpha_1 y_{t-1}^2))^{1/2} a_t$$

where $t = 1, \ldots, n$, $y_0 = 0$, $n = 50, 100, 200$, $\alpha_0 = 0.00001$ and $\alpha_1 = 0.1$, 0.3, 0.5, 0.7, 0.9. The a_t terms are standard normal variables. It is well known that the stationarity conditions for the α coefficients are $\alpha_0 > 0$, $\alpha_1 \geq 0$, and $\alpha_1 < 1$. Here, the choice of α_0 is somewhat arbitrary but is inspired by the case of Engle (1983). In Engle's case, $\alpha_0 = 0.000\,006$. Hence, the choice here is a quantity of comparable magnitude. The series are also simulated 1000 times for each model. The simulated series are fitted as AR(1) models. Both Q^*_{aa} and Q_R are then applied to the residuals of the fitted series to test for ARCH effects.

To compare the robustnesses of Q_R and Q^*_{aa}, the simulations were repeated in Wong and Li (1995) with a_t terms generated from a t distribution with three degrees of freedom. Since a t_3 variable does not possess finite kurtosis, that implicitly means that the time series generated in this fashion will have quite a few outliers. The results are given in Table 6.3.

Table 6.2: *Comparison of empirical p values of Q^*_{aa} and Q_R with outlier(s) (Wong and Li, 1995). Reproduced with the permission of Taylor & Francis Ltd.*

No. of observations (n)	ϕ	No. of outliers	Degrees of freedom (M) 1		4		7		10	
			Q^*_{aa}	Q_R	Q^*_{aa}	Q_R	Q^*_{aa}	Q_R	Q^*_{aa}	Q_R
50	−0.9	3	0.474	0.080	0.214	0.070	0.115	0.067	0.086	0.062
	−0.6		0.109	0.057	0.057	0.052	0.062	0.059	0.049	0.066
	−0.3		0.034	0.047	0.038	0.048	0.048	0.063	0.049	0.062
	0		0.024	0.050	0.042	0.047	0.051	0.054	0.055	0.064
	0.3		0.023	0.053	0.036	0.050	0.041	0.043	0.051	0.061
	0.6		0.102	0.080	0.056	0.068	0.055	0.068	0.058	0.077
	0.9		0.393	0.150	0.220	0.108	0.139	0.093	0.117	0.102
100	−0.9	3	0.690	0.069	0.527	0.060	0.418	0.057	0.308	0.070
	−0.6		0.233	0.060	0.137	0.053	0.109	0.058	0.096	0.069
	−0.3		0.062	0.047	0.058	0.053	0.046	0.060	0.051	0.063
	0		0.035	0.039	0.043	0.047	0.039	0.052	0.046	0.059
	0.3		0.052	0.042	0.043	0.042	0.054	0.048	0.055	0.057
	0.6		0.238	0.056	0.145	0.047	0.116	0.060	0.095	0.058
	0.9		0.650	0.086	0.497	0.065	0.389	0.068	0.296	0.067
200	−0.9	3	0.765	0.063	0.680	0.060	0.621	0.066	0.560	0.065
	−0.6		0.347	0.051	0.216	0.063	0.188	0.069	0.144	0.074
	−0.3		0.073	0.044	0.066	0.047	0.060	0.056	0.050	0.069
	0		0.041	0.046	0.039	0.057	0.052	0.054	0.053	0.053
	0.3		0.065	0.043	0.058	0.052	0.061	0.055	0.061	0.057
	0.6		0.345	0.053	0.250	0.057	0.194	0.056	0.163	0.050
	0.9		0.746	0.065	0.682	0.062	0.631	0.062	0.585	0.063

Table 6.3: *Comparison of the power of the Q^*_{aa} and Q_R statistics under ARCH (Wong and Li, 1995). Reproduced with the permission from Taylor & Francis Ltd.*

No. of observations (n)	α	Distribution	Degrees of freedom (M): 1		4		7		10	
			Q^*_{aa}	Q_R	Q^*_{aa}	Q_R	Q^*_{aa}	Q_R	Q^*_{aa}	Q_R
50	0.1	Standard normal	0.202	0.192	0.104	0.115	0.049	0.103	0.035	0.086
	0.3		0.491	0.379	0.337	0.269	0.177	0.206	0.104	0.182
	0.5		0.584	0.511	0.476	0.413	0.281	0.358	0.192	0.309
	0.7		0.641	0.606	0.536	0.525	0.378	0.458	0.262	0.410
	0.9		0.673	0.689	0.573	0.621	0.432	0.563	0.324	0.526
100	0.1	Standard normal	0.306	0.222	0.200	0.157	0.149	0.130	0.121	0.117
	0.3		0.625	0.519	0.536	0.375	0.433	0.299	0.361	0.264
	0.5		0.757	0.705	0.684	0.577	0.604	0.486	0.523	0.446
	0.7		0.787	0.779	0.754	0.714	0.671	0.650	0.600	0.601
	0.9		0.805	0.831	0.783	0.803	0.719	0.753	0.652	0.728
200	0.1	Standard normal	0.446	0.280	0.312	0.174	0.263	0.153	0.231	0.114
	0.3		0.764	0.702	0.696	0.539	0.632	0.453	0.576	0.410
	0.5		0.837	0.892	0.830	0.802	0.773	0.718	0.732	0.681
	0.7		0.877	0.918	0.865	0.906	0.832	0.863	0.800	0.833
	0.9		0.901	0.916	0.897	0.931	0.863	0.910	0.821	0.901
50	0.1	t	0.309	0.280	0.192	0.156	0.095	0.129	0.054	0.114
	0.3		0.500	0.553	0.361	0.423	0.210	0.352	0.141	0.311
	0.5		0.585	0.644	0.489	0.575	0.313	0.521	0.211	0.473
	0.7		0.594	0.716	0.529	0.686	0.378	0.619	0.277	0.576
	0.9		0.597	0.785	0.537	0.763	0.423	0.707	0.314	0.670
100	0.1	t	0.433	0.384	0.325	0.248	0.264	0.203	0.217	0.191
	0.3		0.664	0.749	0.596	0.617	0.516	0.533	0.448	0.475
	0.5		0.711	0.823	0.674	0.784	0.593	0.725	0.536	0.685
	0.7		0.735	0.859	0.715	0.858	0.653	0.819	0.576	0.782
	0.9		0.749	0.879	0.671	0.909	0.671	0.892	0.615	0.876
200	0.1	t	0.585	0.569	0.466	0.416	0.411	0.342	0.369	0.308
	0.3		0.806	0.915	0.763	0.864	0.693	0.813	0.646	0.759
	0.5		0.835	0.942	0.838	0.941	0.788	0.931	0.741	0.906
	0.7		0.842	0.931	0.861	0.960	0.824	0.945	0.781	0.940
	0.9		0.851	0.920	0.863	0.969	0.830	0.959	0.798	0.951

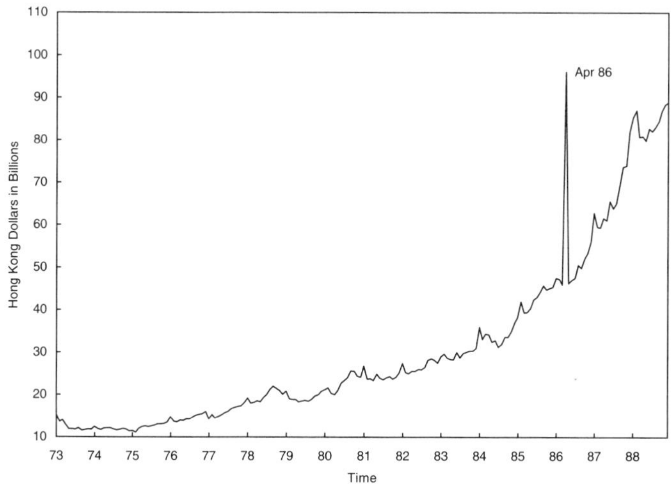

Figure 6.1 *Hong Kong monthly money supply (M1) for 1973–88 (Wong and Li, 1995). Reproduced with the permission of Taylor & Francis Ltd.*

It can be observed that, when the noise is normal, Q^*_{aa} and Q_R are very similar in their power. However, when the noise follows a t_3 distribution, especially for $\alpha_1 = 0.3, 0.5, 0.7, 0.9$, the power of Q_R is always greater than that of Q^*_{aa}. Basically, it can be said that Q_R is uniformally better than Q^*_{aa} in power, in the presence of outliers. For $\alpha_1 = 0.1$, the ARCH model will resemble white noise, which more or less reduces to the situation in Table 6.2. This explains why Q^*_{aa} has better power than that of Q_R in that region.

Example 6.1 *Hong Kong monthly money supply (M1), 1973–88 (Wong and Li, 1995). Reproduced with the permission of Taylor & Francis Ltd.*

Figure 6.1 shows the time plot of the Hong Kong monthly money supply in billions (M1) during the period from January 1973 to December 1988. An outlier clearly appeared in April 1986, because one of the top multinational companies in Hong Kong carried out equity financing during that time.

Figures 6.2 and 6.3 show the autocorrelation function (ACF) and partial ACF (PACF) of the first difference of the M1 series. Standard Box-Jenkins' arguments (Box and Jenkins, 1976) will show that the differenced series is stationary and can be fitted by an MA(1) model. The model is

$$y_t = 385.19 + (1 - 0.65B)a_t \ .$$

It should be noted that the SAS/ETS package is used here for the plotting and fitting of the models.

To understand the structure of the data, Wong and Li (1995) removed and replaced the outlier by a 10-point moving average, and repeated the Box-Jenkins analysis. Figures 6.4 and 6.5 show that the first difference of the series has a mild annual cycle. The ARMA model fitted to the differenced series is

$$(1 - 0.23B^{12})y_t = 372.13 + (1 + 0.58B)a_t \ .$$

It is quite well known that economic data of this type often exhibit some nonlinear behavior. The Ljung-Box test, McLeod-Li test, and the proposed test are applied to the residuals of both series. The degrees of freedom considered are again 1, 4, 7, and 10. The results are summarized in Table 6.4. The smoothed M1 series results indicate clearly that the data contain conditional heteroscedasticity, whereas results for the M1 series show that the rank test detects nonlinearity unambiguously in the presence of outliers, while the other two tests fail.

One suggestion in Wong and Li (1995) is that Q^*_{aa} and Q_R can be used together. If Q^*_{aa} shows no presence of ARCH but Q_R does, then one should be cautioned on the possibility that outliers are present and that, in this situation, the test based on Q_R should be more reliable. When there are ARCH effects and outliers in the data, both the Ljung-Box and McLeod-Li statistics will most probably fail, whereas the Q_R statistic will not. Finally, although Q_R, like Q^*_{aa}, should be most effective in detecting the presence of conditional heteroscedasticity, it can clearly be used to detect other types of nonlinear departure as well.

Table 6.4 *Comparison of three portmanteau statistics (Wong and Li, 1995). Reproduced with the permission of Taylor & Francis Ltd.*

	DF	Q^*_{aa}	Q_R	Ljung-Box
M1	1	3.731	36.789	0.208
	4	4.239	54.565	0.423
	7	4.289	63.148	0.639
	10	4.373	64.400	7.246
Smoothed M1	1	3.705	11.100	1.965
	4	46.464	33.270	4.109
	7	60.835	49.874	6.285
	10	67.523	82.906	10.974

```
Lag Covariance Correlation -1 9 8 7 6 5 4 3 2 1 0 1 2 3 4 5 6 7 8 9 1    Std
------------------------------------------------------------------------------
  0   28559551    1.00000                      |********************  0
  1  -13765694   -0.48200            **********|  .                   0.072357
  2     586424    0.02053                  .   |  .                   0.087569
  3     -62798   -0.00220                  .   |  .                   0.087594
  4     685759    0.02401                  .   |  .                   0.087594
  5    -586422   -0.02053                  .   |  .                   0.087629
  6    -301747   -0.01057                  .   |  .                   0.087654
  7     175945    0.00616                  .   |  .                   0.087661
  8    -761027   -0.02665                  .  *|  .                   0.087663
  9    2521365    0.08828                  .   |** .                  0.087705
 10    -403816   -0.01414                  .   |  .                   0.088169
 11    -589123   -0.02063                  .   |  .                   0.088181
 12    1026902    0.03596                  .   |* .                   0.088207
 13   -1696738   -0.05941                  .  *|  .                   0.088283
 14    3195640    0.11189                  .   |** .                  0.088492
 15   -1393759   -0.04880                  .  *|  .                   0.089230
 16    -628742   -0.02202                  .   |  .                   0.089370
 17     113321    0.00397                  .   |  .                   0.089398
 18     701350    0.02456                  .   |  .                   0.089399
 19   -1692405   -0.05926                  .  *|  .                   0.089434
 20     993584    0.03479                  .   |* .                   0.089640
 21    -100159   -0.00351                  .   |  .                   0.089710
 22    2560855    0.08967                  .   |** .                  0.089711
 23   -2020386   -0.07074                  .  *|  .                   0.090179
 24    1186424    0.04154                  .   |* .                   0.090469
```

Figure 6.2 *Autocorrelation function of first difference of M1 series (Wong and Li, 1995). Reproduced with the permission of Taylor & Francis Ltd.*

```
Lag Correlation -1 9 8 7 6 5 4 3 2 1 0 1 2 3 4 5 6 7 8 9 1
------------------------------------------------------------
  1  -0.48200             **********|  .
  2  -0.27588                 ******|  .
  3  -0.17279                    ***|  .
  4  -0.07685                   .**|  .
  5  -0.05506                   . *|  .
  6  -0.05732                   . *|  .
  7  -0.04361                   . *|  .
  8  -0.07130                   . *|  .
  9   0.05465                   .  |* .
 10   0.09153                   .  |**.
 11   0.06215                   .  |* .
 12   0.09150                   .  |**.
 13   0.00191                   .  |  .
 14   0.12889                   .  |***
 15   0.11747                   .  |**.
 16   0.05714                   .  |* .
 17   0.02793                   .  |* .
 18   0.02655                   .  |* .
 19  -0.06363                   . *|  .
 20  -0.04644                   . *|  .
 21  -0.05137                   . *|  .
 22   0.09680                   .  |**.
 23   0.02929                   .  |* .
 24   0.04492                   .  |* .
```

Figure 6.3 *Partial autocorrelation function of first difference of M1 series (Wong and Li, 1995). Reproduced with the permission of Taylor & Francis Ltd.*

```
Lag Covariance Correlation -1 9 8 7 6 5 4 3 2 1 0 1 2 3 4 5 6 7 8 9 1    Std
--------------------------------------------------------------------------------
  0    2301608     1.00000                     |********************  0
  1     -95811    -0.04163                   . *|  .                 0.072357
  2     229402     0.09967                   .  |**.                 0.072483
  3      18133     0.00788                   .  |  .                 0.073197
  4     178655     0.07762                   .  |**.                 0.073201
  5     126853     0.05511                   .  |* .                 0.073631
  6    -185472    -0.08058                   .**|  .                 0.073847
  7    -114487    -0.04974                   . *|  .                 0.074306
  8     366127     0.15907                   .  |***                 0.074480
  9      91875     0.03992                   .  |* .                 0.076238
 10      50272     0.02184                   .  |  .                 0.076347
 11     287040     0.12471                   .  |**.                 0.076380
 12     553605     0.24053                   .  |*****               0.077439
 13     267905     0.11640                   .  |**.                 0.081256
 14       8056     0.00350                   .  |  .                 0.082124
 15    -104742    -0.04551                   . *|  .                 0.082125
 16     109376     0.04752                   .  |* .                 0.082257
 17     188875     0.08206                   .  |**.                 0.082401
 18      33663     0.01463                   .  |  .                 0.082827
 19    -150860    -0.06555                   . *|  .                 0.082841
 20     121775     0.05291                   .  |* .                 0.083112
 21     -28447    -0.01236                   .  |  .                 0.083288
 22      74764     0.03248                   .  |* .                 0.083298
 23     147868     0.06425                   .  |* .                 0.083364
 24     463161     0.20123                   .  |****                0.083623
```

Figure 6.4 *Autocorrelation function of first difference of smoothed M1 series (Wong and Li, 1995). Reproduced with permission of Taylor & Francis Ltd.*

```
Lag Correlation -1 9 8 7 6 5 4 3 2 1 0 1 2 3 4 5 6 7 8 9 1
------------------------------------------------------------
  1   -0.04163                   . *|  .
  2    0.09811                   .  |**.
  3    0.01588                   .  |  .
  4    0.06956                   .  |* .
  5    0.05953                   .  |* .
  6   -0.09180                   .**|  .
  7   -0.07134                   . *|  .
  8    0.16912                   .  |***
  9    0.06152                   .  |* .
 10    0.00217                   .  |  .
 11    0.13906                   .  |***
 12    0.24177                   .  |*****
 13    0.08936                   .  |**.
 14   -0.01506                   .  |  .
 15   -0.06048                   . *|  .
 16   -0.02328                   .  |  .
 17    0.07339                   .  |* .
 18    0.07965                   .  |**.
 19   -0.06832                   . *|  .
 20   -0.03883                   . *|  .
 21   -0.07748                   .**|  .
 22   -0.02531                   . *|  .
 23    0.05152                   .  |* .
 24    0.17945                   .  |****
```

Figure 6.5 *Partial autocorrelation function of first difference of smoothed M1 series (Wong and Li, 1995). Reproduced with the permission of Taylor & Francis Ltd.*

6.3 Diagnostic checking for ARCH models

A lot has been said in the literature on the modeling of conditional heteroscedastic time series, but not too much work has been done on model checking or model selection for ARCH type models. For example, the asymptotic distribution of the squared residual autocorrelations derived from such models should be useful in checking model adequacy, in particular the specification of the conditional variance h_t. In this regard, the Box-Pierce statistic on the first M-squared standardized residual autocorrelations (denoted $Q^2(M)$) was proposed for checking the adequacy of the different nonlinear ARCH specifications (Higgins and Bera, 1992). However, a χ^2 distribution with M degrees of freedom was used as the large sample distribution for the $Q^2(M)$ statistic. The results of Li and Mak (1994) suggested that this is somewhat misleading. In their paper, a correct portmanteau statistic $Q(M)$ is proposed which is based on the correct large sample distribution of the squared standardized residual autocorrelations. The usefulness of this statistic in modeling nonlinear time series with conditional heteroscedasticity should be similar to that of the Ljung-Box statistic in autoregressive moving average models. (See §2.2.)

Following Li and Mak (1994), let Y_t be a stationary and ergodic time series. Let F_t be the information set (σ-field) generated by all past observations up to and including time t. In practice F_t may contain exogenous random variables as well, but for simplicity it is assumed that F_t is generated by $\{Y_t, Y_{t-1}, \ldots\}$ only. Given F_{t-1}, the distribution of Y_t is assumed to be Gaussian with conditional mean $\mu(\boldsymbol{\theta}; F_{t-1})$ and conditional variance $h(\boldsymbol{\theta}; F_{t-1})$, where $\boldsymbol{\theta}$ is a $l \times 1$ vector of parameters. Let $\mu_t = \mu(\boldsymbol{\theta}; F_{t-1})$ and $h_t = h(\boldsymbol{\theta}; F_{t-1})$ for convenience. Both μ_t and h_t are assumed to be known except for the parameter $\boldsymbol{\theta}$ and they are both assumed to have continuous second-order derivatives almost surely. The above formulation will include Engle's ARCH model as a special case, when $\mu_t = 0$ and $h_t = \alpha_0 + \alpha_1 \epsilon_{t-1}^2 + \cdots + \alpha_r \epsilon_{t-r}^2$. Note also that both μ_t and h_t can be nonlinear functions of past observations. A wide class of h_t has been considered by Higgins and Bera (1992). In practice, $\boldsymbol{\theta}$ would have to satisfy regularity conditions for stationarity and ergodicity but, of course, this will depend on the particular forms of μ_t and h_t. See for instance, Engle and Bollerslev (1986) for a discussion of these conditions for the ARCH and generalized ARCH models. Let $\hat{\boldsymbol{\theta}}$ be the conditional maximum likelihood estimator of $\boldsymbol{\theta}$. Suppose that Y_t is invertible. Let $\epsilon_t = Y_t - \mu_t(\boldsymbol{\theta})$ and $\hat{\epsilon}_t$ be the corresponding residual when $\boldsymbol{\theta}$ is replaced by $\hat{\boldsymbol{\theta}}$. Similarly define $\hat{\mu}_t$ and $\hat{h}_t$. Unlike the homogeneous variance situation ϵ_t, $t = 1, 2, \ldots$, have different conditional variances and the autocorrelation of ϵ_t^2 should take this into account. Similar consideration also applies to the residuals $\hat{\epsilon}_t$. The lag-k-squared (standardized) residual autocorrelation is defined as

$$\tilde{r}_k = \frac{\sum_{t=k+1}^{n} (\hat{\epsilon}_t^2/\hat{h}_t - \bar{\epsilon})(\hat{\epsilon}_{t-k}^2/\hat{h}_{t-k} - \bar{\epsilon})}{\sum_{t=1}^{n} (\hat{\epsilon}_t^2/\hat{h}_t - \bar{\epsilon})^2} \qquad k = 1, 2, \ldots$$

where $\bar{\epsilon} = n^{-1} \sum \hat{\epsilon}_t^2/\hat{h}_t$ and n is the sample size. Since it can be shown that $\bar{\epsilon}$ converges to one in probability if the model is correct, $\tilde{r}_k$ can be replaced by

$$\hat{r}_k = \frac{\sum(\hat{\epsilon}_t^2/\hat{h}_t - 1)(\hat{\epsilon}_{t-k}^2/\hat{h}_{t-k} - 1)}{\sum(\hat{\epsilon}_t^2/\hat{h}_t - 1)^2} .$$

Furthermore, since $n^{-1} \sum(\hat{\epsilon}_t^2/\hat{h}_t - 1)^2$ converges to a constant we need only consider the asymptotic distribution of

$$\hat{C}_k = \frac{1}{n} \sum \left(\frac{\hat{\epsilon}_t^2}{\hat{h}_t} - 1 \right) \left(\frac{\hat{\epsilon}_{t-k}^2}{\hat{h}_{t-k}} - 1 \right) .$$

The result for $\hat{r}_k$ follows immediately by Slustky's thoerem. It can be seen that $\hat{C}_0$ converges to 2 in probability if ϵ_t is Gaussian conditional on F_{t-1}. Denote by C_k the counterpart of $\hat{C}_k$ when $\hat{\epsilon}_t$ and $\hat{h}_t$ are replaced by ϵ_t and h_t, respectively.

First we derive the asymptotic distribution and the information matrix $\boldsymbol{G}$ of $\hat{\boldsymbol{\theta}}$. For each t the contribution to the conditional log-likelihood l by Y_t is, apart from a constant, $l_t = -\frac{1}{2} \log h_t - \frac{1}{2}\epsilon_t^2/h_t$, and $l = \sum l_t$. By direct differentiation,

$$\frac{\partial l}{\partial \boldsymbol{\theta}} = \frac{1}{2} \sum \frac{1}{h_t} \frac{\partial h_t}{\partial \boldsymbol{\theta}} \left(\frac{\epsilon_t^2}{h_t} - 1 \right) + \sum \frac{\epsilon_t}{h_t} \frac{\partial \mu_t}{\partial \boldsymbol{\theta}} . \qquad (6.30)$$

Differentiating again and taking iterative expectations with respect to F_{t-1} (Higgins and Bera, 1992), we have

$$E\left(\frac{\partial^2 l}{\partial \boldsymbol{\theta} \partial \boldsymbol{\theta}^{\mathrm{T}}} \right) = \frac{-1}{2} \sum E\left\{ \frac{1}{h_t^2} \left(\frac{\partial h_t}{\partial \boldsymbol{\theta}} \right) \left(\frac{\partial h_t}{\partial \boldsymbol{\theta}} \right)^{\mathrm{T}} \right\} - \sum E\left\{ \frac{1}{h_t} \frac{\partial \mu_t}{\partial \boldsymbol{\theta}} \left(\frac{\partial \mu_t}{\partial \boldsymbol{\theta}} \right)^{\mathrm{T}} \right\}.$$

Theorem 6.1 *Under the usual regularity conditions (Hall and Heyde, 1980, p.156) for maximum likelihood estimators, $\sqrt{n}(\hat{\boldsymbol{\theta}} - \boldsymbol{\theta})$ is asymptotically normally distributed with mean zero and variance* $\boldsymbol{G}^{-1} = -E(n^{-1} \partial^2 l/\partial \boldsymbol{\theta} \partial \boldsymbol{\theta}^{\mathrm{T}})^{-1}$.

Let $\boldsymbol{C} = (C_1, \ldots, C_M)^{\mathrm{T}}$ and $\hat{\boldsymbol{C}} = (\hat{C}_1, \ldots, \hat{C}_M)^{\mathrm{T}}$, for some integer $M > 0$. Similarly define $\hat{\boldsymbol{r}}$ and $\boldsymbol{r}$. It can be shown, as in McLeod and Li (1983), that $\sqrt{n}\boldsymbol{C}$ is asymptotically normally distributed with mean zero and variance $4 \cdot \mathbf{1}$, where $\mathbf{1}$ is the $M \times M$ identity matrix. Following Li and Mak (1994) a Taylor series expansion of $\hat{\boldsymbol{C}}$ about $\boldsymbol{\theta}$ and evaluated at $\hat{\boldsymbol{\theta}}$ gives

$$\hat{\boldsymbol{C}} \approx \boldsymbol{C} + \frac{\partial \boldsymbol{C}}{\partial \boldsymbol{\theta}}(\hat{\boldsymbol{\theta}} - \boldsymbol{\theta})$$

where $\partial \boldsymbol{C}/\partial \boldsymbol{\theta} = (\partial C_1/\partial \boldsymbol{\theta}, \ldots, \partial C_M/\partial \boldsymbol{\theta})^{\mathrm{T}}$, with

$$\begin{aligned}
\frac{\partial C_k}{\partial \boldsymbol{\theta}} &= n^{-1} \sum \frac{2\epsilon_t}{h_t} \left(\frac{-\partial \mu_t}{\partial \boldsymbol{\theta}} \right) \left(\frac{\epsilon_{t-k}^2}{h_{t-k}} - 1 \right) \\
&\quad - n^{-1} \sum \frac{\epsilon_t^2}{h_t^2} \frac{\partial h_t}{\partial \boldsymbol{\theta}} \left(\frac{\epsilon_{t-k}^2}{h_{t-k}} - 1 \right) \\
&\quad + n^{-1} \sum \left(\frac{\epsilon_t^2}{h_t} - 1 \right) \frac{2\epsilon_{t-k}}{h_{t-k}} \left(- \frac{\partial \mu_{t-k}}{\partial \boldsymbol{\theta}} \right) \\
&\quad + n^{-1} \sum \left(\frac{-\epsilon_{t-k}^2}{h_{t-k}^2} \right) \frac{\partial h_{t-k}}{\partial \boldsymbol{\theta}} \left(\frac{\epsilon_t^2}{h_t} - 1 \right) .
\end{aligned}$$

By the ergodic theorem the first and the last two terms converge to zero in probability and hence for large n,

$$\frac{\partial C_k}{\partial \boldsymbol{\theta}} \approx -\frac{1}{n} \sum \frac{\epsilon_t^2}{h_t^2} \frac{\partial h_t}{\partial \boldsymbol{\theta}} \left(\frac{\epsilon_{t-k}^2}{h_{t-k}} - 1 \right) .$$

By taking expectation with respect to F_{t-1} for each term under the summation sign and by the ergodic theorem, $\partial C_k/\partial \boldsymbol{\theta}$ can be consistently estimated by $X_k = -\sum(1/h_t)(\partial h_t/\partial \boldsymbol{\theta})\{(\epsilon_{t-k}^2/h_{t-k}) - 1\}/n$. However, this quantity does not in general converge to zero since both h_t and $\partial h_t/\partial \boldsymbol{\theta}$ can be correlated with the term in brackets. Define the resultant $M \times l$ matrix by $-\tilde{\boldsymbol{X}}$ when $\partial C_k/\partial \boldsymbol{\theta}$ in $\partial \boldsymbol{C}/\partial \boldsymbol{\theta}$, $k = 1, \ldots, M$, are estimated by $\boldsymbol{X}_k$ and denoting the probability limit of $\tilde{\boldsymbol{X}}$ by $\boldsymbol{X}$, then we have proved the following lemma (Li and Mak, 1994) .

Lemma 6.1 *Under the conditions made earlier in this section,*

$$\hat{\boldsymbol{C}} \approx \boldsymbol{C} - \boldsymbol{X}(\hat{\boldsymbol{\theta}} - \boldsymbol{\theta}) .$$

The vector $\hat{\boldsymbol{C}}$ and hence $\hat{\boldsymbol{r}}$ can be shown to be asymptotically normally distributed by the Mann-Wald device and the martingale central limit theorem (Billingsley, 1961). To obtain the asymptotic covariance of $\hat{\boldsymbol{r}}$ we consider the asymptotic covariance between $\sqrt{n}(\hat{\boldsymbol{\theta}} - \boldsymbol{\theta})$ and $\sqrt{n}\boldsymbol{C}$. Since

$\hat{\boldsymbol{\theta}} - \boldsymbol{\theta} \approx (n\boldsymbol{G})^{-1}\partial l/\partial\boldsymbol{\theta}$, this asymptotic covariance is equal to

$$E\left(\boldsymbol{G}^{-1}\frac{\partial l}{\partial\boldsymbol{\theta}}\boldsymbol{C}^{\mathrm{T}}\right) = \boldsymbol{G}^{-1}E\left(\frac{\partial l}{\partial\boldsymbol{\theta}}\boldsymbol{C}^{\mathrm{T}}\right).$$

From (6.30) the expectation of $\partial l/\partial\boldsymbol{\theta}$ and C_k is equal to

$$n^{-1}E\left[\sum\left\{\frac{1}{2h_t}\frac{\partial h_t}{\partial\boldsymbol{\theta}}\left(\frac{\epsilon_t^2}{h_t}-1\right)+\frac{\epsilon_t}{h_t}\frac{\partial\mu_t}{\partial\boldsymbol{\theta}}\right\}\sum\left(\frac{\epsilon_{t'}^2}{h_{t'}}-1\right)\left(\frac{\epsilon_{t'-k}^2}{h_{t'-k}}-1\right)\right].$$

By taking iterative expectations it can be shown that the cross covariance of $\epsilon_t h_t^{-1}\partial\mu_t/\partial\boldsymbol{\theta}$ and C_k is zero. It can also be seen that

$$E\left\{\frac{1}{h_t}\frac{\partial h_t}{\partial\boldsymbol{\theta}}\left(\frac{\epsilon_t^2}{h_t}-1\right)\left(\frac{\epsilon_{t'}^2}{h_{t'}}-1\right)\left(\frac{\epsilon_{t'-k}^2}{h_{t'-k}}-1\right)\right\}$$

is non-zero if and only if $t' = t$. In which case,

$$\begin{aligned}E\left(\frac{\partial l}{\partial\boldsymbol{\theta}}C_k\right) &= E\left\{(2n)^{-1}\sum h_t^{-1}\frac{\partial h_t}{\partial\boldsymbol{\theta}}\left(\frac{\epsilon_t^2}{h_t}-1\right)^2\left(\frac{\epsilon_{t'-k}^2}{h_{t'-k}}-1\right)\right\}\\ &= E\left\{n^{-1}\sum h_t^{-1}\frac{\partial h_t}{\partial\boldsymbol{\theta}}\left(\frac{\epsilon_{t-k}^2}{h_{t-k}}-1\right)\right\}.\end{aligned}$$

The second equality is obtained by taking conditional expectation of the individual terms with respect to F_{t-1}. Again $E\{(\partial l/\partial\boldsymbol{\theta})C_k\}$ can be consistently estimated by the quantity $n^{-1}\sum(1/h_t)(\partial h_t/\partial\boldsymbol{\theta})\{(\epsilon_{t-k}^2/h_{t-k})-1\}$. Hence, we have proved that the asymptotic cross expectation between $\sqrt{n}(\hat{\boldsymbol{\theta}}-\boldsymbol{\theta})$ and $\sqrt{n}\boldsymbol{C}$ is given by $\boldsymbol{G}^{-1}\boldsymbol{X}^{\mathrm{T}}$. Theorem 6.2 summarizes our discussion above.

Theorem 6.2 *(Li and Mak, 1994)*

$\sqrt{n}\hat{\boldsymbol{r}}$ is asymptotically normally distributed with mean 0 and the asymptotic covariance $\boldsymbol{V}$ is given by $\mathbf{1} - \frac{1}{4}\boldsymbol{X}\boldsymbol{G}^{-1}\boldsymbol{X}^{\mathrm{T}}$.

The result gives more accurate asymptotic standard errors for the squared residual autocorrelations. In practice entries of $\boldsymbol{G}$ can be replaced by the respective sample averages as in Li (1992). An alternative statistic results by replacing the factor $\frac{1}{4}$ with $1/\hat{C}_0^2$. Furthermore,

$$Q(M) = n\hat{\boldsymbol{r}}^{\mathrm{T}}\hat{\boldsymbol{V}}^{-1}\hat{\boldsymbol{r}} \tag{6.31}$$

will be asymptotically χ^2 distributed with M degrees of freedom if the model is correct. This quantity can be used as a statistic for testing the joint significance of $\hat{r}_i$, $i = 1, \ldots, M$. Unlike the Box-Pierce result,

$\boldsymbol{V}$ is in general not idempotent even asymptotically since in general $\frac{1}{4}\boldsymbol{X}^{\mathrm{T}}\boldsymbol{X} \not\approx \boldsymbol{G}$. The matrix $\boldsymbol{V}$ is trivially idempotent if $\partial\mu_t/\partial\boldsymbol{\theta} = 0$, and $(1/h_t)\partial h_t/\partial\boldsymbol{\theta}$ and $\epsilon_{t-k}^2/h_{t-k} - 1$, $k > 0$, are uncorrelated. But this implies that $\boldsymbol{X} = \boldsymbol{0}$ and we have basically the McLeod and Li (1983) result. See section 5.2.

Note that for Engle's autoregressive conditional heteroskedasticity model $\frac{1}{n}\sum\{h_t^{-1}\partial h_t/\partial\boldsymbol{\theta}(\epsilon_{t-k}^2/h_{t-k} - 1)\} \cong 0$ if $k > r$. If $M > k$ then $\boldsymbol{X}$ would have approximately zero entries from the $(r+1)$th row onward. This of course implies that the asymptotic standard errors of $\hat{r}_i$, $i = r+1, \ldots, M$, are just $1/\sqrt{n}$ and that the simplified statistic in Li and Mak (1994),

$$Q(r, M) = n \sum_{i=r+1}^{M} \hat{r}_i^2 \tag{6.32}$$

will be asymptotically χ^2 distributed with $M - r$ degrees of freedom. Hence, $Q(r, M)$ can be used as a portmanteau statistic for testing the overall significance of $\hat{r}_i$, $i = r + 1, \ldots, M$. The result also suggests that the $Q^2(M)$ statistic would in general not be asymptotically χ^2 distributed with M degrees of freedom.

A small simulation experiment is performed in Li and Mak (1994) to assess the usefulness of the asymptotic result obtained. In the experiment, the time series Y_t satisfies the following AR(1)-ARCH(1) model

$$Y_t = \phi_1 Y_{t-1} + \epsilon_t$$

where ϵ_t is normal with mean zero and conditional variance

$$h_t = \alpha_0 + \alpha_1 \epsilon_{t-1}^2 \ .$$

Let $\boldsymbol{\theta} = (\phi_1, \alpha_0, \alpha_1)$. Two sets of parameter values, $\boldsymbol{\theta} = (0.3, 0.3, 0.3)$ and $\boldsymbol{\theta} = (0.6, 0.3, 0.6)$, and four different lengths of realization, namely $n = 60$, 100, 200, and 400, are considered. For each set of model parameters and sample size there are 100 independent replications. The parameter $\boldsymbol{\theta}$ is estimated by conditional maximum likelihood using the Newton-Raphson method with starting value $(0.1, 0.1, 0.1)$. The asymptotic standard errors A_i, $i = 1, \ldots, 6$, of $\hat{r} = (\hat{r}_1, \ldots, \hat{r}_6)$ are obtained from the result in Theorem 6.2. The empirical standard errors S_i of $\hat{r}_i$, $i = 1, \ldots, 6$, are also obtained and are taken to be the "true" standard errors. Table 6.5 presents the empirical standard errors and the averages of the asymptotic standard errors. It can be seen that the asymptotic results match the "true" values quite satisfactorily for n as small as 60. As in the previous sections, the standard error of the lag-one-squared standardized residual autocorrelation is substantially smaller than that given by $1/\sqrt{n}$. The empirical power of the statistics $Q(M)$, $Q(r, M)$, and $Q^2(M)$ were also considered by Li and Mak (1994) using two different data generating processes.

Table 6.5 *The empirical (S_i) and the large sample (A_i) standard errors of squared standardized residual autocorrelations in an AR(1)-ARCH(1) model (Li and Mak, 1994). Reproduced with the permission of Blackwell Publishing*

		i					
		1	2	3	4	5	6
$\theta = (0.3, 0.3, 0.3)$							
$n = 60$	A_i	0.064	0.129	0.129	0.129	0.129	0.129
	S_i	0.058	0.116	0.110	0.118	0.101	0.107
$n = 100$	A_i	0.044	0.100	0.100	0.100	0.100	0.100
	S_i	0.043	0.088	0.086	0.093	0.088	0.096
$n = 200$	A_i	0.033	0.071	0.071	0.071	0.071	0.071
	S_i	0.033	0.075	0.065	0.078	0.061	0.064
$n = 400$	A_i	0.023	0.050	0.050	0.050	0.050	0.050
	S_i	0.023	0.053	0.046	0.051	0.052	0.045
$\theta = (0.6, 0.3, 0.6)$							
$n = 60$	A_i	0.076	0.129	0.129	0.129	0.129	0.129
	S_i	0.067	0.108	0.118	0.128	0.091	0.103
$n = 100$	A_i	0.060	0.100	0.100	0.100	0.100	0.100
	S_i	0.060	0.087	0.090	0.089	0.088	0.087
$n = 200$	A_i	0.044	0.071	0.071	0.071	0.071	0.071
	S_i	0.042	0.061	0.065	0.067	0.061	0.071
$n = 400$	A_i	0.032	0.050	0.050	0.050	0.050	0.050
	S_i	0.032	0.047	0.049	0.050	0.045	0.047

In model (I),

$$Y_t = \phi Y_{t-1} + \epsilon_t$$

with

$$h_t = \alpha_0 + \alpha_1 \epsilon_{t-1}^2 + \alpha_2 \epsilon_{t-2}^2 \,.$$

The parameter values used in the simulation are $\phi = \alpha_0 = \alpha_1 = 0.2$; $\alpha_2 = 0,\ 0.2$; and $n = 100$, 200, and 300. The value of M is 6. Four hundred independent replications are generated for each combination of α_2 and n. The simulated data are estimated assuming an ARCH(1) model for ϵ_t. In model (II) Y_t is again autoregressive of order one but

$$h_t = \alpha_0 + \sum_{i=1}^{5} \alpha_i \epsilon_{t-i}^2 \,.$$

We first set $\phi = \alpha_0 = \alpha_1 = \alpha_2 = 0.2$ and $\alpha_i = 0$ for $i > 2$, and then set $\alpha_3 = 0.1$ and $\alpha_4 = \alpha_5 = 0.05$. The latter case resembles situations with persistence in the conditional variance structure. The generated data are estimated with $r = 2$ and known autoregressive order. The number of replications and the values of n and M are the same as in the first model. The results are summed up in Table 6.6 with entries equal to the proportion of rejections based on the upper 5th percentile of the corresponding asymptotic or presumed χ^2 distributions. The degrees of freedom for $Q(M)$ and $Q^2(M)$ are 6 for all cases. The degrees of freedom for $Q(1,6)$ and $Q(2,6)$ are 5 and 4, respectively. It can be seen that $Q(M)$ has the most reliable sizes in all situations with those of $Q(r,M)$ coming close. In contrast, the statistic $Q^2(M)$ is very conservative in size especially for the second model considered. The powers of $Q(M)$ and $Q(r,M)$ are higher than that of $Q^2(M)$ in all situations. This feature is more prominent in the second model. An interesting observation is that the $Q(r,M)$ statistic in fact comes very close to that of $Q(M)$ in performance. Given its simplicity one may prefer $Q(r,M)$ to $Q(M)$ in checking the adequacy of a fitted ARCH specification.

Table 6.6 *The empirical sizes and power of $Q(M)$, $Q(r,M)$, and $Q^2(M)$. Replications = 400, $M = 6$ (Li and Mak, 1994). Reproduced with the permission of Blackwell Publishing*

	Size			Power		
	$Q(M)$	$Q(r,M)$	$Q^2(M)$	$Q(M)$	$Q(r,M)$	$Q^2(M)$
Model (I), $r = 1$						
$n = 100$	0.048	0.035	0.023	0.158	0.153	0.123
$n = 200$	0.060	0.053	0.035	0.340	0.303	0.258
$n = 300$	0.060	0.050	0.040	0.518	0.508	0.450
Model (II), $r = 2$						
$n = 100$	0.040	0.033	0.010	0.095	0.115	0.060
$n = 200$	0.060	0.025	0.008	0.218	0.208	0.128
$n = 300$	0.053	0.028	0.010	0.363	0.348	0.215

As an illustrative example we consider below the 1980 daily return series of the Hong Kong Hang Seng index. There are 245 observations and the returns R_t are defined as the log differences of the daily closing prices. The sample ACF and PACF of R_t^2 are plotted in Figure 6.6 (Li and Tong, 2001).

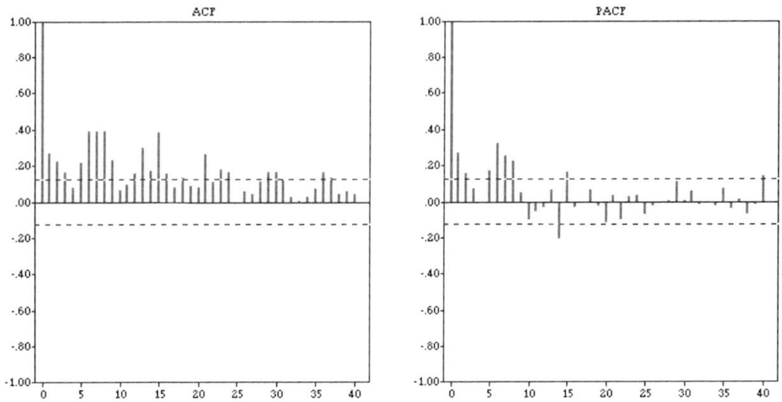

Figure 6.6 *Sample autocorrelations and partial autocorrelation of* R_t^2 *(Li and Tong, 2001). Reproduced with the permission of Elsevier Science*

Example 6.2 *The daily return series of the Hong Kong Hang Seng index, 1980. (Li and Mak, 1994). Reproduced with the permission of Blackwell Publishing.*

We entertain the following model,

$$R_t = \epsilon_t$$

and

$$h_t = \alpha_0 + \sum_{i=1}^{r} \alpha_i \epsilon_{t-i}^2 \ ,$$

which is a slight modification from Li and Mak (1994).

We first consider fitting a model with $r = 5$ *and then a model with* $r = 7$*. Conditional maximum likelihood estimates are obtained using an iteratively weighted least squares scheme as in Mak, Wong, and Li (1997). All estimates have a starting value of 0.1. The major interest is on whether the models fit the data adequately. To this end the first ten* $\hat{r}_k$*s with their large sample standard errors are recorded in Table 6.7. The overall test statistics* $Q(10)$*,* $Q(r, 10)$*, and* $Q^2(10)$ *are also recorded. When* $r = 5$ *both the* $Q(10)$ *and the* $Q(5, 10)$ *statistics clearly reject the model at the upper 5% significance levels of the* χ^2 *distributions with 10 and 5 degrees of freedom, respectively, whereas the* $Q^2(10)$ *statistic suggests that the model is adequate. Note also that* $\hat{r}_5$ *was highly significant using the* *correct large sample standard error. However, it would be insignificant if* $1.96/\sqrt{n}$ *was used as the critical value. Based on the* $\hat{r}_k$*s we fitted a model with* $r = 7$*. In this case, all three* Q *statistics and the individual squared standardized residual autocorrelations suggest an adequate fit to the data. The estimated ARCH model is given by (see also Li and Tong 2001):*

$$\begin{aligned} h_t &= 0.00012 + 0.03997R_{t-1}^2 + 0.13506R_{t-2}^2 + 0.12798R_{t-3}^2 \\ &\quad + 0.15475R_{t-6}^2 + 0.28445R_{t-7}^2 \, . \end{aligned}$$

Table 6.7 *Model diagnostic checking results for the daily return of the Hong Kong Hang Seng Index (1980)* $\hat{r}_k$ *(standard error in parentheses) (Li and Mak, 1994, reproduced with the permission of Blackwell Publishing.)*

k	$r = 5$	$r = 7$
1	−0.0492	−0.0345
	(0.0401)	(0.0222)
2	−0.0202	−0.0266
	(0.0229)	(0.0405)
3	−0.0185	0.0134
	(0.0174)	(0.0392)
4	−0.0471	−0.0500
	(0.0339)	(0.0502)
5	−0.1265	−0.0392
	(0.0450)	(0.0445)
6	0.1289	0.0454
	(0.0639)	(0.0403)
7	0.1602	−0.0351
	(0.0639)	(0.0414)
8	0.0317	0.0718
	(0.0639)	(0.0639)
9	0.0414	−0.0557
	(0.0639)	(0.0639)
10	0.0388	0.0052
	(0.0639)	(0.0639)
$Q(10)$	24.09	13.22
$Q(r, 10)$	11.39	2.03
$Q^2(10)$	16.53	4.17

As suggested earlier the squared standardized residual autocorrelations would be useful tools in checking the adequacy of a conditional heteroscedastic nonlinear time series model. The large sample distribution obtained in this paper clearly enhance their usefulness in applications.

Extension of Theorem 6.2 to the GARCH models were considered by Tse and Zuo (1997) and Ling and Li (1997a). The result in Ling and Li (1997a) applies also to fractionally differenced ARMA processes with GARCH innovations. The major change in the result is in the way that the matrix $\boldsymbol{X}$ is constructed. The $\partial h_t/\partial\boldsymbol{\theta}$ term would have to be evaluated recursively. For example in the GARCH(1, 1) model with $y_t = \epsilon_t$, $h_t = \alpha_0 + \beta h_{t-1} + \alpha_1 \epsilon_{t-1}^2$,

$$
\begin{aligned}
\frac{\partial h_t}{\partial \alpha_0} &= 1 + \beta \cdot \frac{\partial h_{t-1}}{\partial \alpha_0} \\
\frac{\partial h_t}{\partial \alpha_1} &= \frac{\beta \partial h_{t-1}}{\partial \alpha_1} + \epsilon_{t-1}^2
\end{aligned}
\tag{6.33}
$$

and

$$
\frac{\partial h_t}{\partial \beta} = h_{t-1} + \beta \cdot \frac{\partial h_{t-1}}{\partial \beta} .
$$

With appropriate starting values we can evaluate the matrix $\boldsymbol{X}$ as before. Tse and Zuo (1997) also performed more simulation experiments on the Q statistics. Overall the $Q(M)$ statistic using $\mathbf{1} - \frac{1}{4}\boldsymbol{X}\boldsymbol{G}^{-1}\boldsymbol{X}$ works well with Gaussian data while the alternate statistic with 4 replaced by $\hat{C}_0^2$ seems to be the best statistic to employ overall when M is large. Their experiment also suggested that $M = p+q+1$ seems to be a good choice.

6.4 Diagnostics for multivariate ARCH models

Let $\boldsymbol{Y}_t = (y_{1t}, \ldots, y_{kt})^{\mathrm{T}}$ be a k-dimensional time series. The univariate ARCH models can be extended to the k-dimensional case in many ways. This extension began almost as soon as the first paper on ARCH appeared in 1982. See for example, Kraft and Engle (1983), Engle, Granger, and Kraft (1984) and Bollerslev, Engle, and Wooldridge (1988). The extension to multivariate GARCH models resembles that of the multivariate ARMA models. However, among many other things it is necessary to ensure that the multivariate condition covariance matrix $\boldsymbol{V}_t$ is symmetric and positive definite. As in the multivariate ARMA models the number of parameters could grow rapidly with the dimension and the order of the model. In the bivariate case, $\boldsymbol{V}_t (k = 2)$ would have the form

$$
\boldsymbol{V}_t = \begin{pmatrix} h_{11,t} & h_{12,t} \\ h_{21,t} & h_{22,t} \end{pmatrix} . \tag{6.34}
$$

Here $h_{12,t} = h_{21,t}$. Suppose that $\boldsymbol{Y}_t|F_{t-1} \sim N(0, \boldsymbol{V}_t)$. By analogy with the univariate case the first diagonal entry may take the form

$$h_{11,t} = \alpha_{01} + \alpha_{11}y_{1t-1}^2 + \beta_{11}h_{1t-1} + g_{11}(h_{22,t-1}, h_{12,t-1}, y_{2t-1}^2, y_{1t-1}y_{2t-1})$$

where $g_{11}(\ )$ is a linear function of its arguments. There is a similar expression for $h_{22,t}$. The expression for the conditional covariance may assume the form,

$$h_{12,t} = C_{0,12} + C_{12,1}y_{1t-1}y_{2t-1} + b_{12,1}h_{12,t-1} \ .$$

Imposing positive definiteness could be a problem in the multivariate ARCH models. A popular approach is the so-called BEKK representation (Engle and Kroner, 1995). The conditional variance V_t is given by

$$\boldsymbol{V}_t = \boldsymbol{C}_0^{\mathrm{T}}\boldsymbol{C}_0 + \sum_{k=1}^{K}\sum_{i=1}^{q} \boldsymbol{A}_{ik}^{\mathrm{T}} y_{t-i} y_{t-i}^{\mathrm{T}} \boldsymbol{A}_{ik} + \sum_{k=1}^{K}\sum_{i=1}^{p} \boldsymbol{G}_{ik}^{\mathrm{T}} \boldsymbol{V}_{t-i} \boldsymbol{G}_{ik} \ , \quad (6.35)$$

where $\boldsymbol{C}_0$, $\boldsymbol{A}_{ik}$, and $\boldsymbol{G}_{ik}$ are $n \times n$ parameter matrices with $\boldsymbol{C}_0$ triangular; and the summation limit K determines the generality of the process.

It can be shown that representation (6.35) is positive definite under very general conditions. With $K = 1$, $q = 1$ and $p = 0$, $\boldsymbol{A}_{11} = (a_{ij})$, and $\boldsymbol{C}_0^{\mathrm{T}}\boldsymbol{C}_0 = (c_{ij})$ we have,

$$\begin{aligned}
h_{11,t} &= c_{11} + a_{11}^2 y_{1t-1}^2 + 2a_{11}a_{21}y_{1t-1}y_{2t-1} + a_{21}^2 y_{2t-1}^2, \\
h_{12,t} &= c_{12} + a_{11}a_{12}y_{1t-1}^2 + (a_{21}a_{12} + a_{11}a_{22})y_{1t-1}y_{2t-1} \\
&\quad + a_{21}a_{22}y_{2t-1}^2, \\
h_{22,t} &= c_{22} + a_{12}^2 y_{1t-1}^2 + 2a_{12}a_{22}y_{1t-1}y_{2t-1} + a_{22}^2 y_{2t-1}^2 \ .
\end{aligned}$$

Engle and Kroner (1995) contain more details on conditions of stationarity and estimation. The concern of this section is on developing diagnostic checks for the multivariate ARCH models and on checking whether multivariate ARCH models are required. We tackle first the goodness-of-fit problem for multivariate ARCH models by following Ling and Li (1997b).

Let $\{\boldsymbol{Y}_t\}$ be a k-dimensional stationary and ergodic vector time series generated by the equations

$$\begin{aligned}
\boldsymbol{Y}_t - \boldsymbol{\mu}_t &= \boldsymbol{\epsilon}_t \ , \\
\boldsymbol{\epsilon}_t &= \boldsymbol{V}_t^{1/2}\boldsymbol{\eta}_t
\end{aligned} \quad (6.36)$$

where $\mu_t = \mu(\boldsymbol{\theta}, F_{t-1}) = E(\boldsymbol{Y}_t|F_{t-1})$, $\boldsymbol{V}_t = V(\boldsymbol{\theta}, F_{t-1}) = \mathrm{var}(\boldsymbol{Y}_t|F_{t-1})$ is positive definite, and F_t is the σ-field generated by $\{\boldsymbol{Y}_{t-1}, \boldsymbol{Y}_{t-2}, \ldots\}$; $E(\cdot|F_{t-1})$ and $\mathrm{var}(\cdot|F_{t-1})$ denote respectively the conditional expectation given F_{t-1} and the conditional variance given F_{t-1}; $\boldsymbol{\mu}_t$ and $\boldsymbol{V}_t$ are assumed to depend only on F_{t-1} surely (a.s.); $\boldsymbol{V}_t^{1/2}$ is the square root of $\boldsymbol{V}_t$; $\{\boldsymbol{\eta}_t\}$ is a sequence of independent and identically distributed random vectors with mean zero and covariance $\mathbf{1}_k$, where $\mathbf{1}_k$ is the $k \times k$ identity matrix. It is further assumed that $E(\eta_{it}^3) = 0$, $i = 1, \ldots, k$; η_{it} and η_{jt} for $i \neq j$, $j = 1, \ldots, k$, are mutually uncorrelated up to the fourth order and the η_{it}, $i = 1, \ldots, k$, have the same finite fourth-order moment, where η_{it} is the ith component of η_t. The existence of the fourth-order moment is also required in Weiss (1986) for the asymptotic normality of estimators in ARCH models.

Clearly the model (6.36) includes many multivariate linear ARCH errors as a special case. It is a general class of nonlinear multivariate time series models with multivariate ARCH-type errors.

The quasi-conditional log-likelihood l of $\boldsymbol{Y}_1, \boldsymbol{Y}_2, \ldots, \boldsymbol{Y}_n$ is as follows (neglecting a constant):

$$l = \sum_{t=1}^{n} l_t \tag{6.37}$$

and

$$l_t = -\frac{1}{2} \log |\boldsymbol{V}_t| - \frac{1}{2} \boldsymbol{\epsilon}_t^{\mathrm{T}} \boldsymbol{V}_t^{-1} \boldsymbol{\epsilon}_t \tag{6.38}$$

Under the regularity conditions given in Bollerslev and Wooldridge (1992, Theorem 2.1) (see also White, 1994, Theorem 6.2), it can be shown that there exists a sequence of consistent quasi-conditional maximum likelihood estimators $\hat{\boldsymbol{\theta}}$ such that

$$\hat{\boldsymbol{\theta}} - \boldsymbol{\theta} = (n\boldsymbol{B})^{-1} \frac{\partial l}{\partial \boldsymbol{\theta}} + O_p\left(\frac{1}{\sqrt{n}}\right), \quad \sqrt{n}(\hat{\boldsymbol{\theta}} - \boldsymbol{\theta}) \xrightarrow{D} N(0, \boldsymbol{B}^{-1}\boldsymbol{A}\boldsymbol{B}^{-1}) \tag{6.39}$$

where $\xrightarrow{D}$ denotes convergence in distribution, $\boldsymbol{A} = E\{n^{-1}(\partial l/\partial \boldsymbol{\theta})(\partial l/\partial \boldsymbol{\theta})^{\mathrm{T}}\}$, $\boldsymbol{B} = -E(n^{-1}\partial^2 l/\partial \boldsymbol{\theta}/\partial \boldsymbol{\theta}^{\mathrm{T}})$. If η_t follows a multivariate normal distribution, then $\boldsymbol{\theta}$ above is a conditional maximum likelihood estimator and $\boldsymbol{A} = \boldsymbol{B}$. In this case, the asymptotic covariance matrix in (6.39) can be simplified as $\boldsymbol{A}^{-1}$ or $\boldsymbol{B}^{-1}$ (see Bollerslev and Wooldridge, 1992, p.149).

Let $\hat{\boldsymbol{\epsilon}}_t$ be the corresponding residual when the parameter vector $\boldsymbol{\theta}$ in $\boldsymbol{\epsilon}_t$ is replaced by $\hat{\boldsymbol{\theta}}$. Similarly define $\hat{\boldsymbol{\mu}}_t$ and $\hat{\boldsymbol{V}}_t$. The lag l sum of squared (standardized) residual autocorrelations (Ling and Li, 1997b) is defined as

$$\tilde{R}_l = \sum_{t=l+1}^{n} (\hat{\boldsymbol{\epsilon}}_t^{\mathrm{T}} \hat{\boldsymbol{V}}_t^{-1} \hat{\boldsymbol{\epsilon}}_t - \tilde{\boldsymbol{\epsilon}})(\hat{\boldsymbol{\epsilon}}_{t-l}^{\mathrm{T}} \hat{\boldsymbol{V}}_{t-l} \hat{\boldsymbol{\epsilon}}_{t-l} - \tilde{\boldsymbol{\epsilon}}) \Big/ \sum_{t=1}^{n} (\hat{\boldsymbol{\epsilon}}_t^{\mathrm{T}} \hat{\boldsymbol{V}}_t^{-1} \hat{\boldsymbol{\epsilon}}_t - \tilde{\boldsymbol{\epsilon}})^2$$

where $\tilde{\boldsymbol{\epsilon}} = (1/n) \sum_{t=1}^{n} \hat{\boldsymbol{\epsilon}}_t^{\mathrm{T}} \hat{\boldsymbol{V}}_t^{-1} \hat{\boldsymbol{\epsilon}}_t$, $l = 1, 2, \ldots, M$.

If the model is correct then, by the ergodic theorem,

$$\tilde{\boldsymbol{\epsilon}} = \frac{1}{n}\sum_{t=1}^{n} \hat{\boldsymbol{\epsilon}}_t^{\mathrm{T}} \hat{\boldsymbol{V}}_t^{-1} \hat{\boldsymbol{\epsilon}}_t \xrightarrow{\text{a.s.}} E(\boldsymbol{\epsilon}_t^{\mathrm{T}} \boldsymbol{V}_t^{-1} \boldsymbol{\epsilon}_t) \quad \text{as } n \to \infty$$

and, by (6.36), $E(\boldsymbol{\epsilon}_t^{\mathrm{T}} \boldsymbol{V}_t^{-1} \boldsymbol{\epsilon}_t) = E(\boldsymbol{\eta}_t^{\mathrm{T}} \boldsymbol{\eta}_t) = k$. Therefore, for large n $\tilde{R}_l$ can be replaced by

$$\hat{R}_l = \sum_{t=l+1}^{n} (\hat{\boldsymbol{\epsilon}}_t^{\mathrm{T}} \hat{\boldsymbol{V}}_t^{-1} \hat{\boldsymbol{\epsilon}}_t - k)(\hat{\boldsymbol{\epsilon}}_{t-l}^{\mathrm{T}} \hat{\boldsymbol{V}}_{t-l}^{-1} \hat{\boldsymbol{\epsilon}}_{t-l} - k) \Big/ \sum_{t=1}^{n} (\hat{\boldsymbol{\epsilon}}_t^{\mathrm{T}} \hat{\boldsymbol{V}}_t^{-1} \hat{\boldsymbol{\epsilon}}_t - k)^2 .$$

It can be seen that if the model is correct,

$$\frac{1}{n}\sum_{t=1}^{n} (\hat{\boldsymbol{\epsilon}}_t^{\mathrm{T}} \hat{\boldsymbol{V}}_t^{-1} \hat{\boldsymbol{\epsilon}} - k)^2 \xrightarrow{\text{a.s.}} E(\boldsymbol{\epsilon}_t^{\mathrm{T}} \boldsymbol{V}_t^{-1} \boldsymbol{\epsilon}_t - k)^2 \quad \text{as } n \to \infty$$

and

$$E(\boldsymbol{\epsilon}_t^{\mathrm{T}} \boldsymbol{V}_t^{-1} \boldsymbol{\epsilon}_t - k)^2 = E(\boldsymbol{\eta}_t^{\mathrm{T}} \boldsymbol{\eta}_t)^2 - k^2 = \{E(\eta_{it}^4) - 1\}k = ck$$

where $c = E(\eta_{it}^4) - 1$. In particular $c = 2$ if η_t follows the standard multivariate normal distribution. Ling and Li (1997) proposed to use $\hat{R}_l$ as a diagnostic statistic like the $\hat{r}_l$ in ARMA models. To this end they derived the joint asymptotic distribution of $\hat{R}_1, \ldots, \hat{R}_M$.

As in §6.3 we need only consider the asymptotic distribution of

$$\hat{C}_l = \frac{1}{n}\sum_{t=l+1}^{n} (\hat{\boldsymbol{\epsilon}}_t^{\mathrm{T}} \hat{\boldsymbol{V}}_t^{-1} \hat{\boldsymbol{\epsilon}}_t - k)(\hat{\boldsymbol{\epsilon}}_{t-l}^{\mathrm{T}} \hat{\boldsymbol{V}}_{t-l}^{-1} \hat{\boldsymbol{\epsilon}}_{t-l} - k) . \tag{6.40}$$

Let $\boldsymbol{C} = (C_1, C_2, \ldots, C_M)^{\mathrm{T}}$ and $\hat{\boldsymbol{C}} = (\hat{C}_1, \hat{C}_2, \ldots, \hat{C}_M)^{\mathrm{T}}$. Similarly define $\boldsymbol{R}$ and $\hat{\boldsymbol{R}}$.

By the ergodic theorem, it is easy to see that as $n \to \infty$

$$\frac{\partial C_l}{\partial \boldsymbol{\theta}} \xrightarrow{\text{a.s.}} -\boldsymbol{X}_l \tag{6.41}$$

where $\boldsymbol{X}_l = E[(\partial \boldsymbol{V}_t/\partial \boldsymbol{\theta})\text{vec}\{\boldsymbol{V}_t^{-1}(\boldsymbol{\epsilon}_{t-l}^{\mathrm{T}} \boldsymbol{V}_{t-l}^{-1} \boldsymbol{\epsilon}_{t-l} - k)\}]$. Let $\boldsymbol{X} = (\boldsymbol{X}_1, \boldsymbol{X}_2, \ldots, \boldsymbol{X}_M)^{\mathrm{T}}$.

Theorem 6.3 *(Ling and Li, 1997b)*

$$\sqrt{n}\hat{\boldsymbol{C}} \xrightarrow{D} N\{\boldsymbol{O}, (ck)^2 \boldsymbol{\Omega}\} \qquad \text{as } n \to \infty$$

$$\sqrt{n}\hat{\boldsymbol{R}} \xrightarrow{D} N(\boldsymbol{O}, \boldsymbol{\Omega}) \qquad \text{as } n \to \infty$$

where $\boldsymbol{\Omega} = \mathbf{1}_M - \boldsymbol{X}(c\boldsymbol{B}^{-1} - \boldsymbol{B}^{-1}\boldsymbol{A}\boldsymbol{B}^{-1})\boldsymbol{X}^{\mathrm{T}}/(ck)^2$.

The proof is similar in spirit to the proof of theorem (6.2) and is therefore omitted.

From the above theorem, we can obtain more accurate asymptotic standard errors for $\hat{R}_l$, $l = 1, \ldots, M$, and we know as in §6.3 that these asymptotic standard errors are less than $1/\sqrt{n}$ in general. In diagnostic checking, the usual value of $1/\sqrt{n}$ can only be regarded as a crude standard error. However, if $\boldsymbol{V}_t$ is a constant matrix over time, then $\boldsymbol{X} = \boldsymbol{0}$ and the asymptotic standard error of $\hat{R}_l$ is exactly $1/\sqrt{n}$ and hence it will not be affected by the estimate $\hat{\boldsymbol{\theta}}$. If $k = 1$, the special result reduces to that of McLeod and Li (1983).

Like the univariate case, $\boldsymbol{\Omega} = \boldsymbol{1}_M - \boldsymbol{X}(c\boldsymbol{B}^{-1} - \boldsymbol{B}^{-1}\boldsymbol{A}\boldsymbol{B}^{-1})\boldsymbol{X}^{\mathrm{T}}/(ck)^2$ is not an idempotent matrix. Hence, $\hat{\boldsymbol{R}}^{\mathrm{T}}\hat{\boldsymbol{R}}$ is not asymptotically χ^2 distributed. However, the statistic

$$Q(M) = n\hat{\boldsymbol{R}}^{\mathrm{T}}\boldsymbol{\Omega}^{-1}\hat{\boldsymbol{R}} \tag{6.42}$$

will be asymptotically χ^2_M distributed if the model is correct. This quantity should be useful as a portmanteau statistic for checking model adequacy. In practice, $\boldsymbol{X}$, $\boldsymbol{A}$, and $\boldsymbol{B}$ in $\boldsymbol{\Omega}$ can be replaced respectively by the corresponding sample estimates. The constant $(ck)^2$ can be replaced by $\hat{C}_0^2$ and the factor c can be replaced by $\hat{C}_0/k$.

If the multivariate ARCH errors are Kraft and Engle's multivariate linear ARCH errors, i.e., $\mathrm{vec}(\boldsymbol{V}_t) = \alpha_0 + \sum_{i=1}^{r} \alpha_i \,\mathrm{vec}(\boldsymbol{\epsilon}_{t-i}\boldsymbol{\epsilon}_{t-i}^{\mathrm{T}})$, then $\boldsymbol{X}_l$ will be relatively small for $l > r$, and the $(r+1)$th to Mth rows of $\boldsymbol{X}$ will be approximately zero. Thus, the asymptotic standard errors of $\hat{R}_l$, $l = r+1, \ldots, M$, are just $1/\sqrt{n}$ and the statistic

$$Q(r, M) = n \sum_{t=r+1}^{M} \hat{R}_l^2 \sim \chi^2_{M-r} \,. \tag{6.43}$$

Hence, $Q(r, M)$ can be a portmanteau statistic for testing the overall significance of $\hat{R}_l$, $l = r+1, \ldots, M$.

Simulation experiments conducted by Ling and Li (1997b) for some diagonal bivariate ARCH models indicated reasonable size and power properties for $Q(M)$ and $Q(r, M)$. In a more extensive simulation study Tse and Tsui (1999) found that, unlike the Li-Mak test in the univariate case, the multivariate tests have weak power when misspecification occurs in the conditional covariance equations but not in the conditional variances. They also found that an ad hoc Box-Pierce statistic based on the cross-products of standardized residuals has good size and power properties. Let the standardized residuals for the i-th series be

$$\hat{a}_{ti} = \hat{\epsilon}_{ti}/\hat{h}_{iit} \,. \tag{6.44}$$

Let

$$C_{tij} = \begin{cases} \hat{a}_{ti}^2 - 1 & i = j \\ \hat{a}_{ti}\hat{a}_{tj} - \hat{\rho}_{tij} , & i \neq j \end{cases}$$

where $\hat{\rho}_{tij}$ is the conditional correlation $\hat{\rho}_{tij} = \hat{h}_{tij}/(\hat{h}_{tii}\hat{h}_{tjj})^{1/2}$. Denote the lag k autocorrelation of C_{tij} by r_{kij}. Then the proposed Q statistic in Tse and Tsui (1999) is

$$Q(i, j; M) = n \cdot \sum_{k=1}^{M} r_{kij}^2 . \tag{6.45}$$

The reference distribution is χ_M^2 although there is no theoretical justification for the distribution.

Tse (2002) proposed two more residual based tests for diagnostic checking multivariate ARCH models. These were based on the squared standardized residuals $\hat{a}_{ti}^2$ and their cross products C_{tij}. Regressions of C_{tii} and C_{tij}, $i \neq j$, are run on lagged values of $\hat{a}_{ti}^2$, $i = 1, \ldots, k$ and the lagged cross products $\hat{a}_{ti}\hat{a}_{tj}$. Let $\hat{\boldsymbol{d}}_{ti} = (\hat{a}_{t-1,i}^2, \ldots, \hat{a}_{t-M,i}^2)^{\mathrm{T}}$ and $\hat{\boldsymbol{d}}_{tij} = (\hat{a}_{t-1,i}\hat{a}_{t-1,j}, \ldots, \hat{a}_{t-M,i}\hat{a}_{t-M,j})^{\mathrm{T}}$. The following regressions are considered:

$$\begin{aligned} C_{tii} &= \hat{\boldsymbol{d}}_{ti} \cdot \boldsymbol{\delta}_i + \xi_{ti} , & i = 1, \ldots, k , \\ C_{tij} &= \hat{\boldsymbol{d}}_{tij} \cdot \boldsymbol{\delta}_{ij} + \xi_{tij} , & 1 \leq i < j \leq k , \end{aligned} \tag{6.46}$$

where $\boldsymbol{\delta}_i$, $\boldsymbol{\delta}_{ij}$ are vectors of parameters. The asymptotic distribution of the diagnostic tests depends on the following two theorems.

Theorem 6.4 *(Tse, 2002)*

If (6.36) specifies the correct model for the multivariate time series $\{\boldsymbol{Y}_t\}$, then under the regularity conditions of Pierce (1982) $\sqrt{n}\hat{\boldsymbol{\delta}}_i \xrightarrow{D} N(0, \boldsymbol{L}_i^{-1} \boldsymbol{\Omega}_i \boldsymbol{L}_i^{-1})$, where

$$\begin{aligned} \boldsymbol{L}_i &= \operatorname{plim}\left(\frac{1}{n}\sum \boldsymbol{d}_{ti}\boldsymbol{d}_{ti}^{\mathrm{T}}\right) , \\ \boldsymbol{\Omega}_i &= c_i \boldsymbol{L}_i - \boldsymbol{Q}_i \boldsymbol{G} \boldsymbol{Q}_i^{\mathrm{T}} , \end{aligned}$$

with

$$\boldsymbol{Q}_i = \operatorname{plim}\left(\frac{1}{n}\sum \boldsymbol{d}_{ti}\frac{\partial a_{ti}}{\partial \boldsymbol{\theta}^{\mathrm{T}}}\right) ,$$

and $c_i = E\{(a_{ti} - 1)^2\}$, $\boldsymbol{G}$ is the asymptotic covariance matrix of the model parameters $\boldsymbol{\theta}$.

Theorem 6.5 *(Tse, 2002)*

If (6.36) specifies the correct model for the multivariate time series $\{\boldsymbol{Y}_t\}$*, then under the regularity conditions of Pierce (1982)* $\sqrt{n}\hat{\boldsymbol{\delta}}_{ij} \xrightarrow{D} N(0, \boldsymbol{L}_{ij}^{-1} \boldsymbol{\Omega}_{ij}\boldsymbol{L}_{ij}^{-1})$*, where*

$$\boldsymbol{L}_{ij} = \operatorname{plim}\left(\frac{1}{n}\sum \boldsymbol{d}_{tij}\boldsymbol{d}_{tij}^{\mathrm{T}}\right),$$

$$\boldsymbol{\Omega}_{ij} = c_{ij}\boldsymbol{L}_{ij} - \boldsymbol{Q}_{ij}\boldsymbol{G}\boldsymbol{Q}_{ij}^{\mathrm{T}},$$

with

$$\boldsymbol{Q}_{ij} = \operatorname{plim}\left\{\frac{1}{n}\sum \boldsymbol{d}_{tij}\frac{\partial(a_{ti}^2 a_{tj} - \rho_{tij})}{\partial\boldsymbol{\theta}^{\mathrm{T}}}\right\},$$

and $c_{ij} = E\{(a_{ti}^2 a_{tj} - \rho_{tij})^2\}$.

The test statistics are then given by,

$$n \cdot \hat{\boldsymbol{\delta}}_i^{\mathrm{T}} \hat{\boldsymbol{L}}_i \hat{\boldsymbol{\Omega}}_i^{-1} \hat{\boldsymbol{L}}_i \hat{\boldsymbol{\delta}}_i \tag{6.47}$$

and

$$n \cdot \hat{\boldsymbol{\delta}}_{ij}^{\mathrm{T}} \hat{\boldsymbol{L}}_{ij} \hat{\boldsymbol{\Omega}}_{ij}^{-1} \hat{\boldsymbol{L}}_{ij} \hat{\boldsymbol{\delta}}_{ij} \tag{6.48}$$

which are distributed asymptotically as a χ^2_M variable. The hat above represents the corresponding sample estimates.

From the simulation results in Tse (2002) the residual-based diagnostic based on the cross-products of the standardized residual seem to be able to give tests with good size and power.

More recently, Horváth and Kokoszka (2001) and Berks, Horváth, and Kokoszka (2003) gave some further extensions of the results in Li and Mak (1994) and Ling and Li (1997a).

Clearly all the above tests can also be used in testing for no ARCH against the presence of ARCH. For example, for the $Q(M)$ statistic in (6.42) the matrix $\boldsymbol{X}$ is zero if there is no ARCH so that $Q = n\hat{\boldsymbol{R}}^{\mathrm{T}}\hat{\boldsymbol{R}}$ will be asymptotically χ^2_M distributed. Here $\hat{\boldsymbol{R}}$ is constructed from the residuals of the conditional mean model with a constant $\boldsymbol{V}_t = \boldsymbol{V}_0$.

6.5 Testing for causality in the variance

We now turn to two special cases which may be useful in detecting whether multivariate ARCH models are actually needed. Recall in §3.2 the concept of Granger causality was discussed to some detail. The same concept can be usefully applied to study lead-lag relations in the conditional variance of two or more time series. In the financial literature the term volatility is used in place of variance. As in the bivariate ARMA case tests for the presence of lead-lag relationship in the variance between two times series can be derived using the cross-correlations of the squared standardized residuals. Cheung and Ng (1996) were among the first to consider such tests. These tests fall in line with the classical Box-Jenkins framework that is discussed at length in Chapter 3. Therefore, the tests can be easily adapted from existing packages with only slight modifications or transformations on the data. Let the two time series under consideration be denoted by $W_{h,t}$, $h = 1, 2$. In Cheung and Ng (1996) these series are assumed to satisfy stationary ARMA models driven separately by two independent white noise processes a_{1t} and a_{2t}. The causality tests for volatility or the variance process are then based on the squared (ARMA) residuals $\hat{a}_{1t}^2$ and $\hat{a}_{2t}^2$. In reality, instantaneous dependence (causality) often exists among economic or financial time series and it therefore seems appropriate to incorporate this feature into the testing framework. We discuss below this direction of extension which was taken up by Wong and Li (1996). Their result extended that of McLeod (1979) and Haugh (1976).

Following McLeod (1979), let $(W_{1t}, W_{2t})^{\mathrm{T}}$, $-\infty < t < \infty$, be a discrete-time bivariate stationary time series with mean zero. Here $W_{ht} (h = 1, 2)$ can be some suitable differencing of the original series y_{ht} because of stationary requirements. Suppose that W_{ht} can be represented as a univariate stationary and invertible time series of order (p_h, q_h):

$$\phi_h(B) W_{ht} = \theta_h(B) a_{ht} \ , \tag{6.49}$$

where

$$\begin{aligned} \phi_h(B) &= 1 - \phi_{h1}(B) - \cdots - \phi_{hp_h} B^{p_h} \ , \\ \theta_h(B) &= 1 - \theta_{h1}(B) - \cdots - \theta_{hq_h} B^{q_h} \ , \end{aligned}$$

B is the backshift operator, $h = 1, 2$, and a_{1t} and a_{2t} are the individual white-noise series, each marginally having independent and identically distributed terms. It is also assumed that $\{a_{ht}\}$, $h = 1, 2$, have finite eighth moments and are symmetrical about zero. Furthermore, taken together, $\{a_{1t}\}$ and $\{a_{2t}\}$ are assumed to be jointly strictly stationary with finite eighth moments. The innovations $\{a_{ht}\}$, $h = 1, 2$, have mean zero and autocovariance function

$$\begin{aligned} \gamma_{a_h a_h}(l) &= E(a_{ht} a_{h(t+l)}) \\ &= \begin{cases} \sigma_h^2 & \text{if } l = 0, \\ 0 & \text{if } l \neq 0, \end{cases} \end{aligned} \tag{6.50}$$

where σ_h^2 is the individual innovation variance for the time series W_{ht}. The cross-covariance function $\gamma_{a_1a_2}(l)$ and cross-correlation function of a_{1t} and a_{2t} are defined as

$$\gamma_{a_1a_2}(l) = E(a_{1t}a_{2(t+l)})\ , \quad l = 0, \pm 1, \ldots,$$

and

$$\rho_{a_1a_2}(l) = \frac{\gamma_{a_1a_2}(l)}{\sigma_1\sigma_2}\ , \qquad l = 0, \pm 1, \ldots\ .$$

Given n observations W_{ht}, $t = 1, 2, \ldots, n$, from the time series, efficient Gaussian univariate algorithms to estimate the model parameters $\boldsymbol{\beta}_h = (\phi_{h1}, \ldots, \phi_{hp_h}, \theta_{h1}, \ldots, \theta_{hq_h})^{\mathrm{T}}$ have been described by Box and Jenkins (1976), McLeod (1978), and others. The sample innovation cross-covariance and cross-correlation functions of a_{1t} and a_{2t} at lag l are defined by

$$c_{a_1a_2}(l) = n^{-1}\sum_{t=1}^{n-l} a_{1t}a_{2(t+l)}$$

and

$$r_{a_1a_2}(l) = \frac{c_{a_1a_2}(l)}{\{c_{a_1a_1}(0)c_{a_2a_2}(0)\}^{\frac{1}{2}}}\ . \tag{6.51}$$

For any fixed $M \geq 0$, let

$$\boldsymbol{r}_a^{\mathrm{T}} = (r_{a_1a_2}(-M), \ldots, r_{a_1a_2}(-1),\ r_{a_1a_2}(0), r_{a_1a_2}(1), \ldots, r_{a_1a_2}(M))\ ,$$

and

$$\boldsymbol{\rho}_a^{\mathrm{T}} = (\rho_{a_1a_2}(-M), \ldots, \rho_{a_1a_2}(-1),\ \rho_{a_1a_2}(0), \rho_{a_1a_2}(1), \ldots, \rho_{a_1a_2}(M))\ .$$

Let $\boldsymbol{\beta}^{\mathrm{T}} = (\boldsymbol{\beta}_1^{\mathrm{T}}, \boldsymbol{\beta}_2^{\mathrm{T}})$. Denote by $\hat{\boldsymbol{\beta}}_h$ the conditional least-squares estimators of $\boldsymbol{\beta}_h$, $h = 1, 2$. Whittle (1962) obtained an asymptotic distribution of $\hat{\boldsymbol{\beta}}_h$, which depends only on the fourth moments of $\{a_{ht}\}$.

If $\boldsymbol{r}_a$ and $\hat{\boldsymbol{r}}_a$ denote the sample cross-correlation functions for the true and the estimated values of $\boldsymbol{\beta}$, respectively, then $\boldsymbol{r}_a$ and $\hat{\boldsymbol{r}}_a$ are vectors of innovation and residual cross-correlations. McLeod (1979) showed that $\hat{\boldsymbol{r}}_a$ has an asymptotic multivariate normal distribution even when a_{1t} and $a_{2(t+l)}$ are correlated for $l = 0, \pm 1, \ldots, \pm K$.

Let $a_{ht}(h = 1, 2)$ be the innovation series from (6.49). Define $A_{ht} = a_{ht}^2(h = 1, 2)$; then the autocovariance function of the squared residuals can be naturally defined as

$$\begin{aligned}\gamma_{A_hA_h}(l) &= E(A_{ht}A_{h(t+l)}) - E(A_{ht})E(A_{h(t+l)}) \\ &= \begin{cases} \sigma_h^{\prime 2}, & \text{if } l = 0, \\ 0 & \text{if } l \neq 0, \end{cases}\end{aligned} \tag{6.52}$$

The cross-covariance, cross-correlation $\rho_{A_1A_2}(l)$, sample cross-covariance, and sample cross-correlation functions $r_{A_1A_2}(l)$ of A_{1t} and A_{2t} can be defined as in (6.50) and (6.51).

For any fixed $M > 0$, let

$$\boldsymbol{r}^{\mathrm{T}} = (r_{A_1A_2}(-M), \ldots, r_{A_1A_2}(-1),\ r_{A_1A_2}(0), r_{A_1A_2}(1), \ldots, r_{A_1A_2}(M))$$

and

$$\boldsymbol{\rho}^{\mathrm{T}} = (\rho_{A_1A_2}(-M), \ldots, \rho_{A_1A_2}(-1),\ \rho_{A_1A_2}(0), \rho_{A_1A_2}(1), \ldots, \rho_{A_1A_2}(M))\,.$$

Let $\hat{\boldsymbol{r}}$ be the counterpart of $\boldsymbol{r}$ using the squared residuals $\hat{a}^2_{ht}$. It is assumed that a_{1t} and $a_{2(t+l)}$ are independent for $l < -K$ or $l > K$, for some $K > 0$. Elements of $\boldsymbol{\rho}$ and the variance of $\boldsymbol{r}$ are assumed to be finite.

Many economic time series are known to be contemporaneously correlated. Suppose the two time series satisfy the relationship

$$\rho_{A_1A_2}(0) = \rho' \neq 0\,, \quad \rho_{A_1A_2}(l) = 0\,, \quad \text{for} \quad l \neq 0\,.$$

This condition can be interpreted as instantaneous causality of volatility between the two series. We state the following result which is essentially in Wong and Li (1996), without proof.

Theorem 6.6 *Under instantaneous causality only and with the conditions of symmetry for a_{1t} and a_{2t}, $\sqrt{n}\hat{\boldsymbol{r}}$ is asymptotically normal with mean vector*

$$\sqrt{n}(\underbrace{0, \ldots, 0}_{M},\ \rho',\ \underbrace{0, \ldots, 0}_{M})^{\mathrm{T}}$$

and the covariance matrix $\boldsymbol{E}$ which is a diagonal matrix with ones on the main diagonal except at the $(M+1)$th entry.

Naturally, instantaneous causality in volatility is a common phenomenon for many economic and financial time series. This result extends that of Cheung and Ng (1996) to this important situation. Note that this result is also simpler than the result stated in (3.15). It is instructive to consider an example illustrating stochastic processes which are marginally white noise, but have nontrivial lagged dependence in the squared processes. Let a_{1t}, a_{2t}, and a_{3t} be three zero-mean, constant-variance, independent, and identically distributed sequences. The three sequences are also mutually independent. Now consider as in Wong and Li (1996),

$$\begin{aligned} X_{1t} &= a_{1t} + a_{2t}\,, \\ X_{2t} &= \rho a_{1t} + \sqrt{\alpha_0 + \alpha_1 a^2_{2(t-1)}}\, a_{3t}\,, \end{aligned}$$

where $0 < |\rho| < 1$, and α_0 and α_1 are positive constants. The following properties are evident:

(1) Marginally X_{1t} and X_{2t} are both white noise sequences.

(2) X_{1t} and X_{2t} have instantaneous causality.

(3) $E(X_{1t}X_{2t'}) = 0$ if $t \neq t'$.

(4) X_{1t}^2 and X_{2t}^2 have nonzero correlations at both lag 1 and lag 0.

Clearly more examples can be constructed along similar lines.

Let

$$\hat{\boldsymbol{r}}^{(1)} = (\hat{r}_{A_1A_2}(-M), \ldots, \hat{r}_{A_1A_2}(-1))^{\mathrm{T}} ,$$
$$\hat{\boldsymbol{r}}^{(2)} = (\hat{r}_{A_1A_2}(1), \ldots, \hat{r}_{A_1A_2}(M))^{\mathrm{T}} .$$

To test the null hypothesis

$$H_0^{(1)} : \rho_{A_1A_2}(-1) = \cdots = \rho_{A_1A_2}(-M) = 0$$

or

$$H_0^{(2)} : \rho_{A_1A_2}(1) = \cdots = \rho_{A_1A_2}(M) = 0$$

against their simple negations, respectively, under $\rho_{A_1A_2}(0) \neq 0$, the following statistics are proposed:

$$\hat{Q}_M^{(h)} = n(\hat{\boldsymbol{r}}^{(h)})^{\mathrm{T}}(\hat{\boldsymbol{r}}^{(h)}) , \quad h = 1, 2 .$$

Now from Theorem 6.6, it is clear that $\hat{Q}_M^{(h)}$ will follow a χ^2 distribution with M degrees of freedom when both null hypotheses $H_0^{(1)}$ and $H_0^{(2)}$ are true.

As in Chapter 2 the performance of $\hat{Q}_M^{(h)}$ can be improved in finite samples by

$$\tilde{Q}_M^{(h)} = \hat{Q}_M^{(h)} + \frac{M(M+1)}{2n} , \quad h = 1, 2 .$$

Example 6.3 *The S & P 500 and the Toronto stock-exchange index. (Wong and Li, 1996). Reproduced with the permission of the Statistical Society of Canada.*

As reported in Cheung and Ng (1996), because of their theoretical and practical importance, national stock-market indices are widely studied by economists and statisticians. In this case, Standand & Poor's 500 Composite index (S & P 500) and the Toronto stock-exchange index were studied by using our squared-residuals test. Data collected were the daily closing prices of the two indices between January 3, 1989 and June 28, 1991, a span of two and a half years and a total of 630 observations. Let X_{1t} *and* X_{2t} *represent the logarithm of the S & P 500 and the Toronto index, respectively. It was found that the S & P 500 index, after first-differencing, was a white noise series. This is probably a very well-known fact. The Toronto index follows an AR(1) model after first-differencing. Letting* $W_{2t} = X_{2t} - X_{2(t-1)}$, *the model is*

$$W_{2t} = 0.255 W_{2(t-1)} + a_{2t} \ .$$

Cross-correlation of the residuals, from $r_{a_1 a_2}(-20)$ *to* $r_{a_1 a_2}(20)$ *are plotted in Figure 6.7 using the SAS/ETS package. Using the traditional two-standard-error band at the 5% level here (i.e.,* ± 0.079*), other than* $r_{a_1 a_2}(0)$, *only* $r_{a_1 a_2}(-1)$ *and* $r_{a_1 a_2}(-10)$ *are found to be marginally significant. Now if we look at cross-correlations of the squared residuals, the picture is completely different. Values of* $r_{A_1 A_2}(-20)$ *to* $r_{A_1 A_2}(20)$ *are shown in Figure 6.8. Other than* $r_{A_1 A_2}(0)$, $r_{A_1 A_2}(1)$ *is very significant using the conventional error band. The last statement can be justified by our portmanteau test* $\hat{Q}_M^{(1)}$ *and* $\hat{Q}_M^{(2)}$. *This is because* $\hat{Q}_5^{(2)}(= 31.55)$, $\hat{Q}_{10}^{(2)}(= 33.41)$, $\hat{Q}_{20}^{(2)}(= 38.13)$ *are all significant, whereas* $\hat{Q}_M^{(1)}$ *is significant only for* $M = 1$.

Two points are particularly noteworthy. The high significance of $\hat{Q}_M^{(2)}$ *gives strong evidence that the S & P 500 index leads the Toronto index in variance, which concurs with the stock-market wisdom. Another interesting point is that* $r_{a_1 a_2}(1)$ *is clearly nonsignificant, whereas* $r_{A_1 A_2}(1)$ *is highly significant. This demonstrates that the Box-Jenkins model is able to capture the linear structure of the two innovation series but not the second-order structure.*

Another point that is worth mentioning is that when the same analysis was applied to the original data rather than the logged data, the results were almost identical. Also, instead of the S & P 500 index, results with the Dow Jones Industrial Average were very similar. This gave further evidence that the cross-correlation tests are quite robust. In fact, this example and the result for the empirical size with heavy-tailed distributions in Wong and Li (1996) gives us confidence in applying the statistics to stock returns. The cross-correlation tests were applied to several other international stock indices, and similar patterns were observed.

Finally, from the simulations and the last example, it can be seen that a plot of the cross-correlations of the squared residuals, together with the $\hat{Q}_M^{(h)}$ *statistics, provides a set of tools useful in detecting nonlinearity of the innovations. The test is probably most sensitive in detecting nonlinearity involving second order moments. The results may also be useful in the understanding of causality of volatilities between different financial time series.*

```
           -1.0 -0.8 -0.6 -0.4 -0.2  0.0  0.2  0.4  0.6  0.8  1.0
             +----+----+----+----+----+----+----+----+----+----+
-20  -0.041                          *|
-19   0.045                           |*
-18   0.022                           |*
-17   0.005                           |
-16  -0.062                         **|
-15  -0.074                         **|
-14   0.047                           |*
-13   0.007                           |
-12   0.007                           |
-11   0.017                           |
-10   0.086                           |**
 -9  -0.014                           |
 -8   0.019                           |
 -7  -0.030                          *|
 -6  -0.053                          *|
 -5  -0.023                          *|
 -4  -0.027                          *|
 -3  -0.002                           |
 -2   0.019                           |
 -1   0.071                           |**
  0   0.692                           |*****************
  1   0.009                           |
  2   0.069                           |**
  3   0.001                           |
  4   0.037                           |*
  5   0.039                           |*
  6  -0.018                           |
  7   0.011                           |
  8  -0.058                          *|
  9   0.021                           |*
 10   0.059                           |*
 11  -0.001                           |
 12   0.016                           |
 13   0.008                           |
 14   0.053                           |*
 15  -0.007                           |
 16   0.019                           |
 17   0.016                           |
 18   0.024                           |*
 19   0.011                           |
 20   0.010                           |
```

Figure 6.7 *Residual cross-correlation of S & P's 500 and the Toronto Stock Exchange index (Wong and Li, 1996). Reproduced with the permission of the Statistical Society of Canada*

```
              -1.0 -0.8 -0.6 -0.4 -0.2  0.0  0.2  0.4  0.6  0.8  1.0
               +----+----+----+----+----+----+----+----+----+----+
-20  -0.042                            *|
-19   0.006                             |
-18  -0.002                             |
-17  -0.016                             |
-16  -0.004                             |
-15  -0.018                             |
-14  -0.002                             |
-13  -0.010                             |
-12  -0.010                             |
-11  -0.016                             |
-10   0.002                             |
 -9   0.004                             |
 -8  -0.004                             |
 -7  -0.016                             |
 -6  -0.012                             |
 -5  -0.005                             |
 -4   0.013                             |
 -3  -0.016                             |
 -2  -0.006                             |
 -1   0.077                             |**
  0   0.800                             |********************
  1   0.219                             |*****
  2   0.024                             |*
  3   0.007                             |
  4  -0.006                             |
  5   0.023                             |*
  6   0.028                             |*
  7  -0.045                            *|
  8   0.005                             |
  9  -0.014                             |
 10  -0.012                             |
 11  -0.036                            *|
 12   0.013                             |
 13   0.026                             |*
 14  -0.021                            *|
 15  -0.038                            *|
 16  -0.029                            *|
 17  -0.011                             |
 18  -0.030                            *|
 19  -0.027                            *|
 20  -0.033                            *|
```

Figure 6.8 *Cross-correlations of squared residuals of S & P's 500 and the Toronto Stock Exchange index (Wong and Li, 1996). Reproduced with the permission of the Statistical Society of Canada*

CHAPTER 7

Fractionally differenced process

7.1 Introduction

In the last two decades there has been considerable interest in time series models with longer "memory" than those of the autoregressive moving average (ARMA) type. By long-memory it is meant that the autocovariance function γ_k of the process has a much slower decaying rate than those of the usual stationary time series models. For instance, one way to achieve longer memory is to allow $\sum_{k=-\infty}^{\infty} |\gamma_k|$ to be divergent. Long-memory models appear in economics, finance, hydrology, and climatology. For example, in economics, Granger (1980b) has shown that long-memory models can arise from aggregating simple dynamic micro-relationships. More recently, Ding, Granger, and Engle (1993) and Granger, Spear and Ding (2000) suggested that the absolute returns of daily data for a number of financial series exhibit the long memory property. Booth, Kaen and Koveos (1982) and Cheung (1993) suggested that long memory structure may be present in some exchange rate series. Cheung and Lai (1993) studied purchasing power parity using the long memory concept. Baillie (1996) gave a comprehensive review on financial applications of long memory time series. In climatology, tree-ring width variation in trees are being used to backcast climatological patterns several hundreds of years before the first scientific record (LaMarche, 1974). In hydrology, long-memory time series models have long been a subject of interest and are closely related with the Hurst phenomenon, (Lawrance and Kottegoda, 1977; Hipel and McLeod, 1978). See Beran (1994) for more examples.

One particular type of long-memory model can be obtained by considering the operator $\nabla^d = (1 - B)^d$ where B is the backshift operator $BX_t = X_{t-1}$ and d does not necessarily take on integral values. For all d, the power series expansion of $(1 - Z)^d$ exists for $|Z| < 1$, hence if d is not integral valued $(1 - B)^d$ is given by the power series expansion

$$1 - dB - \frac{1}{2}d(1-d)B^2 - \frac{1}{6}d(1-d)(2-d)B^3 \cdots . \qquad (7.1)$$

Given that $\phi(B) = 1 - \phi_1 B - \cdots - \phi_p B^p$ and $\theta(B) = 1 - \theta_1 B \cdots - \theta_q B^q$ satisfying the condition that all their roots are outside the unit circle and $|d| < \frac{1}{2}$, it has been shown by Hosking (1981) that the second order moments of the process X_t defined by

$$\phi(B)\nabla^d X_t = \theta(B)a_t \tag{7.2}$$

exist, where a_t is, assumed to be a sequence of independent identically distributed $(0, \sigma^2)$ variates. The process X_t defined by (7.2) will be called the fractional autoregressive integrated moving average, FARIMA(p, d, q) process. These processes are thus natural generalizations of the mixed autoregressive moving average processes. Unlike the integrated processes where d takes on only integral values, these ARMA(p, d, q) processes are stationary with finite variances. For the case $p = q = 0$, Granger and Joyeux (1980) appeared to be the first to introduce such models.

Hosking (1981) extended the FARIMA$(0, d, 0)$ models to the general FARIMA (p, d, q) cases. For second order properties of the process, it can be shown that the spectral density $f(\lambda)$ of (7.2) is given by

$$\begin{aligned} f(\lambda) &= \frac{\theta(B)\theta(F)}{\phi(B)\phi(F)}[(1 - B)\cdot(1 - F)]^{-d} \\ &= \frac{\theta(B)\theta(F)}{\phi(B)\phi(F)}\left(2\sin\frac{1}{2}\lambda\right)^{-2d}, \quad 0 < \lambda \le \pi , \end{aligned} \tag{7.3}$$

where $F = B^{-1}$.

Thus the spectral density explodes at $\lambda = 0$ and hence the spectral distribution cannot be obtained from the integral of $f(\lambda)$.

If $p = q = 0$ in (7.2) the autocovariance function γ_k of $\{X_t\}$ is of the form

$$\gamma_k = \frac{(-1)^k(-2d)!}{(k - d)!(-k - d)!}, \tag{7.4}$$

where $d! = \Gamma(d + 1)$ and $\Gamma(x)$ is the Gamma function.

Note that for $k = 0, 1, \ldots,$

$$\rho_k = \gamma_k/\gamma_0 = \frac{(-d)!(k + d - 1)!}{(d - 1)!(k - d)!}$$

and for k large enough $\rho_k \sim \frac{(-d)!}{(d-1)!}k^{2d-1}$ and the partial autocorrelation function ϕ_{kk} assumes the simple form

$$\phi_{kk} = \frac{d}{k - d}, \tag{7.5}$$

which suggests a moment estimate of d based on the sample partial autocorrelations. Chong (2000) developed this approach of estimating d further.

For the general case the autocovariance functions and the partial autocorrelations are much more complicated in form. The autocovariances γ_k is given by

$$\gamma_k = \sum_{j=-\infty}^{\infty} \gamma_j^u \cdot \gamma_{k-j}^x \tag{7.6}$$

where γ_{k-j}^x is given by (7.4) and γ_j^u is the usual autocovariance function of an ARMA(p, q) process. Alternatively γ_k can be written in terms of the hypergeometric functions $F(\cdot)$ as in Hosking (1981),

$$\begin{aligned}\gamma_k &= \sum_{j=-q}^{q} \gamma_j^u \gamma_{k-j}^x + \gamma_{k+q}^x \sum_{l=1}^{p} \alpha_l \{F(1, d-q-k; 1-d-q-k; \epsilon_l) - 1\} \\ &\quad + \gamma_{k-q}^x \sum_{l=1}^{p} \alpha_\alpha \{F(1, k-q+d; k-q+1-d; \epsilon_l) - 1\} \,, \end{aligned} \tag{7.7}$$

where ϵ_l arises from the solution of the pth order difference equation

$$\phi(B)\gamma_k = 0 \,.$$

Consider the purely fractional differenced process

$$\nabla^d X_t = a_t \,.$$

For the range of values of d, $0 \le d < \frac{1}{2}$, the autocorrelations and partial autocorrelations are all positive indicating long-term persistency. For the range of values $-\frac{1}{2} < d < 0$, the autocorrelations and partial autocorrelations are all negative and are termed anti-persistent after Mandelbrot (1977). The quantity $H = d + \frac{1}{2}$ is called the Hurst coefficient. Chapter 10 of Hipel and McLeod (1994) gave a detailed treatment of the so-called Hurst phenomenon in hydrology.

Hosking (1996) gives the asymptotic distributions of the sample mean, autocovariance, and autocorrelation of some general long-memory process characterized by autocovariance $\gamma_k \sim ck^{-\alpha}$.

7.2 Methods of estimation

Note that the spectral density $f(\lambda)$ of a FARIMA(p, d, q) process has the form

$$f(\lambda) = \left(2 \sin \frac{1}{2}\lambda\right)^{-2d} \cdot g(\lambda) \tag{7.8}$$

where $g(\lambda)$ is the spectral density of the ARMA(p, q) process $W_t = \nabla^d X_t$. Taking logarithms in (7.8) gives a linear relationship between $\ln f(\lambda)$ and $\ln|1 - e^{-i\lambda}|^2$, i.e.,

$$\ln f(\lambda) = \ln g(0) - d\ln\left(2\sin\frac{1}{2}\lambda\right)^2 + \ln|g(\lambda)/g(0)| \qquad (7.9)$$

Geweke and Porter-Hudak (1983) suggested an estimator $\tilde{d}$ of d by regressing the log-periodogram $I(\lambda_j)$, an estimate of $\ln f(\lambda_j)$, on $\ln(2\sin\frac{1}{2}\lambda_j)^2$, where $\lambda_j = 2\pi j/n \in (0, \pi)$. This provides a quick but not efficient way of estimating d. See Beran (1994, Ch.3) for a discussion on this approach. In what follows, we concentrate on asymptotically efficient procedures. Such procedures have been considered by Li (1981), Li and McLeod (1986), Fox and Taqqu (1986), Beran (1984), Hosking (1984), Dahlhaus (1989), Sowell (1992), and Yajima (1985).

7.2.1 The exact likelihood procedure

Let $\dot{\boldsymbol{\beta}}^{\mathrm{T}} = (\dot{\phi}_1, \ldots, \dot{\phi}_p, \dot{\theta}_1, \ldots, \dot{\theta}_q, \dot{d}, \dot{\sigma}^2) = (\dot{\beta}_1, \ldots, \dot{\beta}_r)$ where $r = p+q+2$ and the dot is used to indicate values of $\boldsymbol{\beta}$ satisfying the assumptions of (7.2). Let a_t be Gaussian white noise.

For a realization of X_t with length n and Gaussian a_t, the exact likelihood can be written

$$L(\boldsymbol{\beta}|\boldsymbol{X}) = (2\pi)^{-\frac{1}{2}n}|\boldsymbol{\Omega}|^{-\frac{1}{2}} \exp -\frac{1}{2}\boldsymbol{X}^{\mathrm{T}}\boldsymbol{\Omega}^{-1}\boldsymbol{X} \qquad (7.10)$$

where $\boldsymbol{X}^{\mathrm{T}} = (X_1, X_2, \ldots, X_n)$ and $\boldsymbol{\Omega}$ is the covariance matrix of the process X_t. The likelihood can be evaluated using a method similar to that of McLeod and Hipel (1978) and Ansley (1979). The approach is to transform X_t into a set of independent identically distributed zero mean random variables $\boldsymbol{e}^{\mathrm{T}} = (e_1, e_2, \ldots, e_n)$ with $\mathrm{var}(e_i) = \sigma^2$. The transformation matrix $\boldsymbol{L}$ may be obtained from the Cholesky decomposition of $\boldsymbol{\Omega}$. The method proposed here, however, is slightly different in the method of orthogonalizing $\boldsymbol{X}$. Durbin's recursive formula (Box and Jenkins, 1976, p.82) is used rather than the Cholesky decomposition. The sample autocovariances are replaced by the theoretical autocovariances when $\boldsymbol{\beta} = \dot{\boldsymbol{\beta}}$ in Durbin's formula. The independent variables e_i can be shown to be identical to those obtained via the Cholesky decomposition method. The latter approach is, however, particularly appealing in the purely fractional differenced process because the kth partial autocorrelation function assumes the simple form $d/(k - d)$. Thus the regression coefficients of X_t on past X's can be easily obtained by using only half of Durbin's algorithm (Box and Jenkins, 1976, equation A3.2.7).

Following Li (1981), define

$$\tilde{e}_t = X_t - X(t|t-1)$$

where $X(t|t-1) = E(X_t|X_s, t > s \geq 1)$, and the $\tilde{e}_t$ are independent and normally distributed variates by the projection theorem (Loeve, 1978, p.127). Let the $(t-1)$th prediction error $\sigma^2(t|t-1)$ relative to σ^2 be

$$\sigma^2(t|t-1) = E(\tilde{e}_t^2)/\sigma^2 \tag{7.11}$$

Now making the change of variable using

$$\tilde{e}_t = X_t - X(t|t-1)$$

since the Jacobian is 1, the logarithm of the likelihood (7.10) is given by (Schweppe, 1965; Brockwell and Davis, 1991, §8.7)

$$\begin{aligned} \log L(\boldsymbol{B}|X) &= \text{constant} - \frac{1}{2}\sum \log \sigma^2(t|t-1) - \frac{n}{2}\log \sigma^2 \\ &\quad -\frac{1}{2}\sum \tilde{e}_t^2 \Big/ [\sigma^2\sigma^2(t|t-1)] \; . \end{aligned} \tag{7.12}$$

Let $e_t = \tilde{e}_t/\sigma(t|t-1)$ then e_t are independent normal variates with mean 0 and variance σ^2. Since $\boldsymbol{X}$ is Gaussian $X(t|t-1)$ is given by the regression equation

$$X(t|t-1) = \phi_{t-1,1}X_{t-1} + \phi_{t-1,2}X_{t-2} + \cdots + \phi_{t-1,t-1}X_1 \; , \tag{7.13}$$

where the $\phi_{t-1,j}$'s can be computed from Durbin's algorithm. The $t-1$th prediction error would then be given by

$$\sigma^2(t|t-1) = \sigma^2(t-1|t-2)(1-\phi_{t-1,t-1}^2) \; , \tag{7.14}$$

or recursively,

$$\begin{aligned} \sigma^2(t|t-1) &= \sigma^2(t-1|t-2)(1-\phi_{t,t}^2) \\ &= \sigma_X^2(1-\phi_{t-1,t-1}^2)(1-\phi_{t-2,t-2}^2)\cdots(1-\phi_{1,1}^2) \; . \end{aligned} \tag{7.15}$$

For a purely fractional differenced process where $p = q = 0$, σ_X^2 is just $(-2d)!/(-d!)^2$ and can be calculated easily from the power series expansion of the gamma function. If $p \neq 0$ or $q \neq 0$ then the autocovariance function can be computed using (7.7);

$$\gamma_k = \sum_{j=-\infty}^{\infty} \gamma_j^u \gamma_{k-j}^x \; ,$$

where γ_j^u, γ_{k-j}^x are defined in (7.7). Thus, the likelihood function can be evaluated exactly (apart from a truncation error in computing γ_k which can be made arbitrarily small). The full Durbin's algorithm will, of course, have to be used if p or $q \neq 0$.

Maximizing (7.12) over σ^2 gives

$$(\log L)_{\max} = \text{constant} - \frac{1}{2} n \log S - \frac{1}{2} \sum \log \sigma^2(t|t-1)$$

where $S = \sum e_t^2$.

The log-likelihood is now concentrated only on $\boldsymbol{\eta} = (\phi_1, \ldots, \phi_p, \theta_1, \ldots, \theta_q, d)^{\mathrm{T}}$. A nonlinear optimization algorithm may then be used to obtain maximum likelihood estimates of $\boldsymbol{\beta}$.

7.2.2 An approximate maximum likelihood procedure

The exact likelihood procedure is appealing but computer time consuming and quickly becomes very complicated when the values of p and q increase. It would then be of practical importance if an approximate maximum likelihood procedure (Box and Jenkins, 1976) could be used to obtain an estimate of $\boldsymbol{\beta}$. Moreover, in reality, very few processes will have really infinite memory. It seems, therefore, reasonable and perhaps realistic during estimation or in the assumed model, to approximate ∇^d by a sufficiently long truncation of its power series expansion. Given that the process is actually governed by (7.2),

$$\phi(B)\nabla^d X_t = \theta(B) a_t$$

and that can be approximated by

$$\dot{a}_t = \dot{p}(B) \sum_{j=1}^{q} \dot{\theta}_j \dot{a}_{t-j} + \sum_{i=1}^{p} \dot{\phi}_i X_{t-i} \tag{7.16}$$

where $\dot{p}(B)$ is a polynomial of degree k obtained by truncating $\nabla^{-d} \cdot \dot{p}(B)$ and can be written as

$$1 + \dot{\psi}_1 + \dot{\psi}_2 B^2 + \cdots + \dot{\psi}_k B^k$$

where

$$\dot{\psi}_i = \frac{(i + \dot{d} - 1)!}{i!(\dot{d} - 1)!} \ .$$

It follows from the Kakeya-Enström Theorem (Henrici, 1974, p.462) that $\dot{p}(B) = 0$ has all roots outside the unit circle. Hence, (7.16) is stationary for any $k \geq 0$. Moreover, for k large enough and d not larger than $\frac{1}{2}$ the difference in models (7.2) and (7.16) can be made negligible.

Given $\dot{a}_t$ and the assumption of normality we can evaluate the approximate log-likelihood of $\dot{\boldsymbol{\beta}}$ as

$$\log L \cong \text{constant} - \frac{1}{2} n \log \dot{\sigma}^2 - \frac{1}{2\dot{\sigma}^2} \sum_{t=1}^{n} \dot{a}_t^2 \ . \tag{7.17}$$

Note that

$$\begin{aligned}\frac{\partial a_t}{\partial d} &= (\ln\nabla)a_t \\ &= -\Big(B + \frac{B^2}{2} + \frac{B^3}{3} + \cdots\Big)a_t \\ &= \delta_{t-1} \ . \end{aligned} \tag{7.18}$$

It can be shown that δ_{t-1} is stationary with finite fourth order moment. This result will be useful in deriving asymptotic properties of the estimator $\hat{\boldsymbol{\beta}}$ and in deriving diagnostic tests.

Hosking (1984) considered a similar approach by considering the truncation

$$\begin{aligned}\nabla^d X_t &\cong \nabla_M^d X_t \\ &= \sum_{j=0}^{t+M-1} \pi_j X_{t-j} \ .\end{aligned}$$

A small simulation in Hosking (1984) suggested that an $M = 30$ gave very reasonable estimates. Obviously the value of M can be allowed to increase with the sample size n. The backcasting method of Box and Jenkins (1976, Ch.7) can be adapted to the FARIMA(p, d, q) models following McLeod and Holanda Sales (1983). The backward and forward equations are

$$\begin{aligned} p(B)X_t = b_t \ , &\quad \phi(B)b_t = \theta(B)a_t \\ p(F)X_t = c_t \ , &\quad \phi(F)c_t = \theta(B)e_t \end{aligned}$$

where $F = B^{-1}$ and e_t is a sequence of independent normal variables with mean zero and variance σ^2.

Let $\hat{\boldsymbol{\beta}}$ be an asymptotically efficient estimator of $\boldsymbol{\beta}$. As shown in Li and McLeod (1986) and Li (1981), $\sqrt{n}(\hat{\boldsymbol{\beta}} - \boldsymbol{\beta})$ is asymptotically normal with information matrix $\mathbf{I}$ given by

$$\frac{1}{\sigma^2}\begin{bmatrix} \boldsymbol{H} & \vdots & \boldsymbol{J} & \vdots & 0 \\ \cdots & \cdots & \cdots & \cdots & \cdots \\ \boldsymbol{J}^{\mathrm{T}} & \vdots & \frac{\pi^2}{6}\sigma^2 & \vdots & 0 \\ \cdots & \cdots & \cdots & \cdots & \cdots \\ 0 & \vdots & 0 & \vdots & \frac{1}{(2\sigma^2)} \end{bmatrix}$$

where

$$\boldsymbol{J}^{\mathrm{T}}_{1(p+q)} = [(\gamma_{\delta u}(i-1)), (\gamma_{\delta v}(i-1))] \ ,$$

where

$$\gamma_{\delta v}(i-1) = \sigma^2 \sum_{j=1}^{\infty} \frac{1}{i+j} \theta'_j ,$$

and

$$\gamma_{\delta u}(i-1) = \sigma^2 \sum_{j=1}^{\infty} \frac{1}{i+j} \phi'_j .$$

$\sigma^{-2}\boldsymbol{H}$ is the information matrix for the usual ARMA(p,q) process. It may be noted, surprisingly, that the variance of $\hat{d}$ does not depend on d!. In fact if $p = q = 0$, $\text{var}(\hat{d}) = \pi^2/(6n)$. Thus the information matrix of $\hat{\boldsymbol{\eta}} = (\hat{\phi}_1, \ldots, \hat{\phi}_p, \ldots, \hat{\theta}_1, \ldots, \hat{\theta}_q, \hat{d})^{\mathrm{T}}$ is

$$\bar{\boldsymbol{H}} = \frac{1}{\sigma^2} \begin{bmatrix} \boldsymbol{H} & \vdots & \boldsymbol{J} \\ \cdots & \cdots & \cdots \\ \boldsymbol{J}^{\mathrm{T}} & \vdots & \dfrac{\pi^2}{6}\sigma^2 \end{bmatrix} . \tag{7.19}$$

7.3 A model diagnostic statistic

Let $\boldsymbol{\eta}$ be the population analog of $\hat{\boldsymbol{\eta}}$ and $\hat{a}_t$ be the residuals resulting from fitting the FARIMA(p,d,q) model. As before let $\hat{r}_k$ be the lag k residual autocorrelation.

As shown in Li (1981) the joint asymptotic distribution of $\sqrt{n}(\hat{\boldsymbol{\eta}} - \boldsymbol{\eta}, \hat{\boldsymbol{r}})$ where $\hat{\boldsymbol{r}}^{\mathrm{T}} = (\hat{r}_1, \ldots, \hat{r}_m)$ is normal with mean 0 and covariance matrix

$$\left[\begin{array}{c|c} \bar{\boldsymbol{H}}^{-1} & -\bar{\boldsymbol{H}}^{-1}\boldsymbol{X}^T \\ \hline -\boldsymbol{X}\bar{\boldsymbol{H}}^{-1} & \mathbf{1}_m \end{array}\right] ,$$

where $\bar{H}$ is the information matrix defined in (7.19) for $\boldsymbol{\eta} = (\phi_1, \ldots, \phi_p, \theta_1, \ldots, \theta_q, d)$ and

$$\boldsymbol{X} = \begin{bmatrix} & \vdots & 1 \\ & \vdots & 1/2 \\ \boldsymbol{Y} & \vdots & \vdots \\ & \vdots & 1/m \end{bmatrix}$$

where $\boldsymbol{Y} = (-\phi'_{i-j} | \theta'_{i-j})_{m \times (p+q)}$ as in (2.7).

The following result is obtained as in previous chapters.

Theorem 7.1 $\sqrt{n} \cdot \hat{\boldsymbol{r}}$ *is asymptotically normal with covariance matrix*

$$\mathbf{1}_m - \boldsymbol{X}\bar{\boldsymbol{H}}^{-1}\boldsymbol{X}^{\mathrm{T}}$$

It is easily seen that for m sufficiently large $\boldsymbol{X}^{\mathrm{T}}\boldsymbol{X} = \bar{\boldsymbol{H}}$ and thus $\mathbf{1}_m - \boldsymbol{X}\bar{\boldsymbol{H}}^{-1}\boldsymbol{X}^{\mathrm{T}}$ is approximately idempotent with rank $m-p-q-1$. This implies at once that as (2.11) for $n \gg m$ large enough

$$Q_m = n \cdot \sum_{l=1}^{m} \hat{r}_a^2(l)$$

is approximately $\chi^2(m-p-q-1)$ distributed. A portmanteau type statistic can thus be defined in a similar way as in the ARMA(p,q) case. However, as in other cases Q_m may be very conservative in the ARMA(p,q) case; some modification of Q_m is usually required in actual practice. A modified portmanteau statistic is given by the $\tilde{Q}_m$ statistic in (2.12)

$$\tilde{Q}_m = n(n+2)\sum_{l=1}^{m} \hat{r}_a^2(l)/(n-l) \ .$$

Now in practice, $\hat{a}_t$ would be approximated by $\hat{a}_t^*$ which is obtained by a truncation of ∇^d. On the other hand, if the exact likelihood procedure is used, a set of n prediction errors e_t, $t = 1, \ldots, n$, are produced by the algorithm. Therefore, it is more convenient to use the e_t's in model diagnostic checking. Li (1981) argued that as in the ARMA(p,q) case (Ansley, 1981) $\boldsymbol{r}_e = (r_e(1), \ldots, r_e(m))^{\mathrm{T}}$ has the same asymptotic distribution as $\hat{\boldsymbol{r}}$, where $r_e(k)$ is the lag-k autocorrelation of e_t.

It can be seen that the variance of $\hat{r}_a(k)$ in an ARMA(p,d,q) process, where $|d| < \frac{1}{2}$, does not depend on the value of d. For $p = q = 0$, the matrix $\boldsymbol{X}$ is just the vector $\left(1, \frac{1}{2}, \ldots, \frac{1}{m}\right)^{\mathrm{T}}$, hence, for sufficiently large m

$$\boldsymbol{X}(\boldsymbol{X}^{\mathrm{T}}\boldsymbol{X})^{-1}\boldsymbol{X}^{\mathrm{T}} = \frac{6}{\pi^2}\begin{bmatrix} 1, & \frac{1}{2}, & \frac{1}{3}, & \cdots, & \frac{1}{m} \\ \frac{1}{2} & \frac{1}{2.2} & \frac{1}{3.3}, & & \\ \vdots & & & \ddots & \\ \cdots & \cdots & \cdots & \cdots & \frac{1}{m.m} \end{bmatrix} \tag{7.20}$$

Thus the variance of $\hat{r}_a(k)$ is given by,

$$\frac{1}{n}\cdot\left(1 - \frac{6}{k^2\pi^2}\right) .$$

The variance rapidly approaches $\frac{1}{n}$ as k increase. For $p = 1$, that is,

$\phi_1 \neq 0$ and $q = 0$, the situation is more complicated. $\boldsymbol{X}$ is now given by

$$\begin{bmatrix} 1 & 1 \\ \phi & 1/2 \\ \vdots & \vdots \\ \phi^{m-1} & 1/m \end{bmatrix} \tag{7.21}$$

and $\boldsymbol{X}^{\mathrm{T}}\boldsymbol{X}$ is easily seen to be

$$\begin{bmatrix} \sum_{i=1}^{m} \phi^{2i} & \sum_{i=1}^{m} \frac{1}{i}\phi^{i-1} \\ \sum_{i=1}^{m} \frac{1}{i}\phi^{i-1} & \sum_{i=1}^{m} \frac{1}{i^2} \end{bmatrix}.$$

Now if $\phi \neq 0$, $|\phi| < 1$, then

$$\begin{aligned} \lim_{m\to\infty} \sum_{i=1}^{m} \frac{1}{i}\phi^{i-1} &= \sum_{i=1}^{\infty} \frac{1}{i}\phi^{i-1} \\ &= \frac{-\ln(1-\phi)}{\phi}. \end{aligned}$$

Consequently, the asymptotic information matrix is given by

$$\begin{bmatrix} \frac{1}{1-\phi^2} & \frac{-\ln(1-\phi)}{\phi} \\ \frac{-\ln(1-\phi)}{\phi} & \frac{\pi^2}{6} \end{bmatrix}. \tag{7.22}$$

After some algebra the asymptotic variance of $\hat{r}_a(k)$ is found to be

$$\frac{1}{n}\left[1 - \frac{1}{\Delta}\left(\frac{1}{k^2(1-\phi^2)} + \frac{2\cdot\phi^{k-1}\ln(1-\phi)}{k\phi} + \frac{\pi^2}{6}\cdot\phi^{2k-2}\right)\right],$$

where $k \geq 1$ and Δ is the determinant of (7.22).

Simulation experiments have been performed to test the validity of the results in Theorem 7.1. Only the purely fractional differenced processes for $0 \leq d < \frac{1}{2}$ are considered because it is for this range of d that applications are most likely to occur. The series length for each replication is 250 and the values of d are $0, .1, .2, .3$, and $.4$, respectively. In the first experiment the fractionally differenced processes are generated exactly using the partial autocorrelations, (7.5) and (7.15). The random number generator Super-duper (Marsaglia, 1976) was used together with the Box-Muller method to generate the e_t's. The exact likelihood procedure is then used to estimate d. There are 400 replications for each value of d

chosen. In the second experiment truncated processes are simulated and d is estimated using the unconditional least squares method. The truncated process consists of the first 50 terms of the power series expansion of ∇^d. There are 500 replications for each d. Tables 7.1 and 7.2 summarize up the results of the two experiments respectively. The number of rejections of the portmanteau statistics $\tilde{Q}_m$, at the upper 5% level of the chi-square $m-1$ distribution, for $m = 20$, are recorded in the second column. The third and fourth columns record the sample mean and standard deviation of the portmanteau statistics. The sample standard deviations of the residual autocorrelation at lag 1 are also recorded in the last column.

Table 7.1 *Empirical significance of the portmanteau test using exact likelihood*

d	Number of rejections at 5%	$\tilde{Q}_m$	$SD(Q_m^*)$	$SD(r_e(1))$
0	28	19.51	6.780	.0401
.1	25	18.97	6.592	.0389
.2	22	18.71	6.304	.0410
.3	21	19.05	6.367	.0400
.4	30	19.40	6.965	.0441

$n = 250$, $m = 20$, number of replications = 400.
Exact procedure.

Table 7.2 *Empirical significance of the portmanteau test using unconditional least squares. (Li and McLeod, 1986). ©1986 Biometrika Trust, reproduced with the permission of Oxford University Press*

d	Number of rejections at 5%	$\tilde{Q}_m$	$SD(Q_n^*)$	$SD(\hat{r}_a(1))$
0	25	19.12	6.354	.0394
.1	23	18.92	6.275	.0389
.2	19	18.91	6.584	.0384
.3	23	18.37	6.416	.0394
.4	27	18.91	6.433	.0412

$n = 250$, $m = 20$, number of replications = 500.
Simulations truncated after 50th term of ∇^{-d}.

It can be seen that the mean and standard deviation of $\tilde{Q}_m$ in both experiments are very close to the mean and variance of a $\chi^2(19)$ variate. The sample standard deviation of the first residual autocorrelation is also very close to the theoretical value of $\frac{1}{\sqrt{n}}\sqrt{(1-6/\pi^2)} = .0396$. The number of rejections performs fairly well for the first experiment but not as good as that of the second.

Other goodness-of-fit tests have been considered in the literature. Robinson (1991) considered testing for dynamic conditional heteroscedasticity and/or serial correlation when the underlying process is long-memory in moments of order 2, 3, and 4. Beran (1992) considered testing the null hypothesis $H_0 : f(\lambda) = f(\lambda;\theta)$ where $f(\lambda)$ is the spectral density of X_t, against the alternative $H_1 : f(\lambda) \neq f(\lambda;\theta)$ when X_t could be long-memory. The statistic is based on comparing the periodogram $I(\lambda_j)$ of X_t with $f(\lambda_j;\theta)$ (Milhoj, 1981) and can be written in the form

$$(2\pi)^{-1}\sum_{k=0}^{n-1}(\hat{\gamma}_k/\hat{\gamma}_0)^2$$

where

$$\hat{\gamma}_k = 4\pi n^{-1}\sum_{j=1}^{n^*} I(\omega_j)/f\{\omega_j\}\cos(k\omega_j)\ , \quad k = 1,\ldots,n-1\ ,$$

$\omega_j = 2\pi j/n$, $j = 1, 2, \ldots, n^*$, $n^* = (n-1)/2 - \frac{1}{2}$ if $n-1$ is odd and $n^* = (n-1)/2$ if $n-1$ is even, are the estimated covariances of the residual process $\hat{a}_t$ arising from fitting a general linear model to X_t. The form bears some resemblance to the Q_m statistic (2.11).

Example 7.1 *As an application, an* $ARMA(0, d, 0)$ *model is constructed for the logarithmic transformed tree ring width indices (1700–1960) taken from the upper treeline of Campito Mountain, California (courtesy of Dr. V.C. LaMarche, Jr., the University of Arizona). This series is of considerable climatological interest. The sample autocorrelations and partial autocorrelations are displayed in Figure 7.1. It can be seen that all the sample autocorrelations Figure 7.2 are positive and decay in an approximately hyperbolic manner. In addition, most of the partial autocorrelations are also positive. As a result, this series would often be considered nonstationary although an AR(4) model with* $\phi_3 = 0$ *can be constructed. Nevertheless, since the sample autocorrelations decay in hyperbolic fashion, the alternative* $ARMA(p, d, q)$ *model with* $0 < |d| < \frac{1}{2}$ *can also be considered. For simplicity only, an* $ARMA(0, d, 0)$ *is fitted to the series. The exact likelihood procedure is used and the estimate of d is found to be 0.4275 with a standard error of .0794. The value of the portmanteau statistic at* $m = 20$ *is* 12.67 *indicating no lack of fit. The residual autocorrelations and their 95% confidence limits for the* $ARMA(4, 0, 0)$

model and the fractional differenced model are shown in Tables 7.3 and 7.4, respectively. Although in both cases the first residual autocorrelation slightly exceeds the 95% limits, the overall pattern indicates whiteness of the residuals and the portmanteau statistics in both cases are small (13.86 and 12.67, respectively).

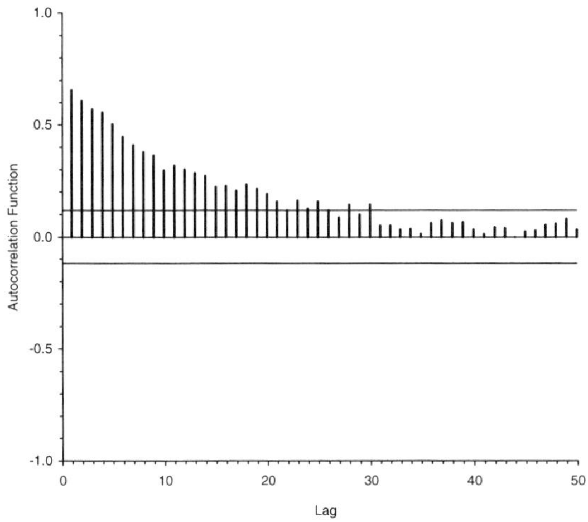

Figure 7.1 *Sample autocorrelations of the tree-ring data*

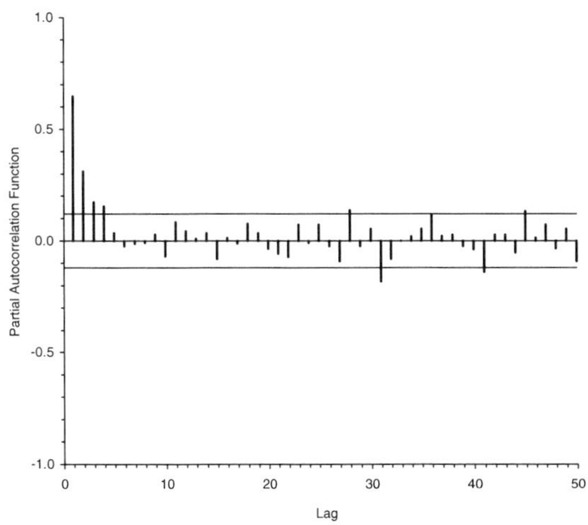

Figure 7.2 *Partial autocorrelation of the tree-ring data*

Table 7.3 *Residual autocorrelations $ARMA(4,0,0)$ model*

	Residual autocorrelations	95% confidence limit
1	−0.046560	0.041807
2	−0.058790	0.053665
3	0.074190	0.112112
4	−0.059470	0.080909
5	0.011600	0.114895
6	0.003700	0.115620
7	−0.030070	0.118247
8	−0.005340	0.116581
9	0.038610	0.118541
10	−0.105760	0.118698
11	0.054260	0.119482
12	0.037010	0.119521
13	0.018100	0.119932
14	0.063430	0.120089
15	−0.062480	0.120344
16	−0.004800	0.120481
17	−0.020920	0.120638
18	0.065370	0.120736
19	0.051130	0.120834
20	0.048730	0.120912

Although the residual variance is not much less than that of an ARMA $(4,0,0)$ model ($\cong 4 \times 10^{-2}$ in both cases) there is only 1 parameter in the frational differenced model while there are three in the $ARMA(4,0,0)$ model ($\phi_3 = 0$). The fractional differenced model is thus more parsimonious in terms of the number of estimated parameters than that of the $ARMA(4,0,0)$ model. The FARIMA model provides an alternate competing model.

Table 7.4 *Residual autocorrelations fractional differenced model*

	Residual autocorrelations	95% confidence limit
1	−0.091150	0.075966
2	0.022840	0.111722
3	0.010590	0.117152
4	0.090350	0.118994
5	0.033850	0.119837
6	−0.000190	0.120292
7	−0.013330	0.120566
8	−0.010520	0.120743
9	0.041920	0.120865
10	−0.107870	0.120952
11	0.026050	0.121016
12	0.013430	0.121065
13	0.004880	0.121103
14	0.042440	0.121133
15	−0.075700	0.121157
16	−0.001900	0.121177
17	−0.048550	0.121193
18	0.051320	0.121207
19	0.028220	0.121219
20	0.019640	0.121229

7.4 Diagnostics for fractional differencing

Agiakloglou and Newbold (1994) considered two diagnostic tests for testing ARMA(p, q) models against FARIMA(p, d, q) models. Recall that in (7.18) $\partial a_t / \partial d = -a_{t-1} - a_{t-2}/2 - a_{t-3}/3 - \cdots$. Hence the score with respect to d is

$$\left.\frac{\partial \log L}{\partial d}\right|_{d=0} = -\frac{1}{\hat{\sigma}^2} \sum_t \sum_k \frac{1}{k} \hat{a}_{t-k} \hat{a}_t$$

where the residuals $\hat{a}_t$ are from an ARMA(p, q) model fitted to the series X_t. This suggests an LM type test statistic based on the score

$$S_m = \sum_{k=1}^{m} \frac{1}{k} \hat{r}_k \, , \tag{7.23}$$

for some integer m.

Agiakloglou and Newbold (1994) proposed two methods to compute the LM tests. The t-test is based on the regression

$$\hat{a}_t = \sum_{i=1}^{p} \beta_i W_{t-i} + \sum_{j=1}^{q} \gamma_j Z_{t-j} + \delta K_m + u_t$$

where

$$K_m = \sum_{k=1}^{m} \frac{1}{k} \hat{a}_{t-k} \, , \quad \hat{\theta}(B) W_t = X_t, \quad \hat{\theta}(B) Z_t = \hat{a}_t \, . \tag{7.24}$$

The test statistic is the usual t-test for $\delta = 0$.

The Z-test is directly based on S_m. It can be shown that

$$\text{var}(S_m) = \boldsymbol{h}' \boldsymbol{W} \boldsymbol{h}$$

where

$$\boldsymbol{W} = (n+2)^{-1} \cdot \boldsymbol{LCL} \, ,$$

where $\boldsymbol{C}$ is var$(\hat{\boldsymbol{r}})$, which is given by (2.8), and L is an $m \times m$ diagonal matrix with i-th diagonal element $(n - i)^{1/2}$ and $\boldsymbol{h}$ is an $m \times 1$ vector with kth element k^{-1}. The test statistic is

$$Z = (\boldsymbol{h} \hat{\boldsymbol{W}} \boldsymbol{h})^{-1/2} S_m$$

where $\hat{W}$ is evaluated using the fitted ARMA(p, q) model. Under the null hypothesis that $d = 0$, Z is asymptotically standard normal. The authors showed by some simulation that the t-test is more powerful with negative d, while the Z-test is more powerful with a positive d. Note that all the results in this chapter assume that the process mean is zero. It was shown in Hosking (1982) and Samarov and Taqqu (1988) that if the mean μ of X_t is unknown and is estimated by either the maximum likelihood method or the sample mean, then $\hat{\mu}$ has variance of the order n^{2d-1}. Fortunately Dahlhaus (1989) showed that the asymptotic distribution of $\hat{\boldsymbol{\beta}}$ remains the same whether the mean μ is known or estimated. This is consistent with the simulation result reported in Li and McLeod (1986). However, as demonstrated in Agiakloglou and Newbold (1994) the effect of estimating the mean μ could be conspicuous if the sample size is small.

variance and the sampling variance of $\hat{r}_1$ for ARMA(1, 0) models, when $\phi_1 = 0, 0.2, 0.4, 0.6$, and 0.8 with variances of the innovations equal to 1. The length of each series and the number of replications for each values of ϕ_1 was taken as 100. The results are summed up in Table 8.1. Values inside the bracket are two times the standard error of the empirical variance of $\hat{r}_1$. It can be seen that the theoretical and sampling variances are in reasonable agreement.

Table 8.1 *Empirical variance of $\hat{r}_1$ for autoregressive process of order 1*

ϕ_1	Theoretical variance of $\hat{r}(1)$	Empirical variance of $\hat{r}(1)$
0	.0093	.0104 (±.0030)
0.2	.0095	.0087 (±.0022)
0.4	.0097	.0113 (±.0039)
0.6	.0098	.0094 (±.0024)
0.8	.0100	.0084 (±.0035)

Series length = 100
Number of replications = 100
$\text{Var}(a_t) = 1$.

8.2 Other non-Gaussian time series

Much attention has been paid to the construction of time series with pre-specified marginal distributions. For example, Lawrance and Lewis (1977; 1985) considered models with exponential marginals. An exponential MA(1) model can be constructed as follows:

$$X_t = \begin{cases} p\,a_t & \text{with probability } p \\ p\,a_t + a_{t+1} & \text{with probability } 1-p\,, \end{cases}$$

whereas an AR(1) process with exponential marginals can be defined by (Gaver and Lewis, 1980),

$$X_t = \begin{cases} p\,X_{t-1} & \text{with probability } p \\ p\,X_{t-1} + E_i & \text{with probability } 1-p\,. \end{cases}$$

where $\{E_i\}$ is an i.i.d. sequence of exponential random variables with parameter λ. McKenzie (1985) considered a collection of simple models for discrete valued time series. See also Jacobs and Lewis (1978a, b). Smith (1986) raised some concerns on the estimation of this kind of models in practice.

A more fruitful route has been taken by Zeger and Qaqish (1988) and Li (1991, 1994). Motivated by biomedical applications Zeger and Qaqish (1988) proposed the so-called Markov regression models by extending the idea of generalized linear models (McCullagh and Nelder, 1989). This is essentially a conditional likelihood approach and seems reasonable if one is relatively sure about the structure of the conditional mean and variance of a process $\{y_t\}$. As in the i.i.d. case, these models are able to handle processes with constant coefficient of variation and overdispersion. Another advantage is that quite reasonable estimates of model parameters can usually be obtained by using the method based on quasi-likelihood and iteratively weighted least squares. Li (1991) considered model diagnostic checking for this type of model. Zeger and Qaqish (1988) used the bootstrap to evaluate the goodness-of-fit of one of their examples, as the asymptotic distribution of the residual autocorrelations was then unknown. This asymptotic distribution has been derived by Li (1991) which facilitates model diagnostic checking. In addition, an easy to use score statistic was derived and was shown to have reasonable performance in checking model adequacy. The residual autocorrelations can be used as a supplement to the score statistic in checking the adequacy of a model. In this connection, Jung and Tremayne (2003) considered tests for serial dependence in time series models of counts. The models considered here are examples of the so-called observational driven models of Cox (1981). Zeger (1988) considered a parameter driven model.

Let $\{y_t\}$ be the time series process under consideration. Let X_t be a $p \times 1$ vector of covariates. Let F_t be the information set $\{X_t, \ldots, X_1, y_{t-1}, \ldots, y_1\}$. The conditional mean and variance of $\{y_t\}$ given F_t are denoted by μ_t and $V(\mu_t)\phi$, respectively. It is assumed that

$$g(\mu_t) = X_t'\beta + \sum_{i=1}^{q} \theta_i f_i(F_t) , \tag{8.6}$$

where g is called the link function, β is a vector of parameters, and $\{f_i\}$ are functions of past observations (Zeger and Qaqish, 1988). For canonical links, $\partial g/\partial \mu = 1/V(\mu)$. Let $\theta^{\mathrm{T}} = (\theta_1, \ldots, \theta_q)$ and $\gamma^{\mathrm{T}} = (\beta^{\mathrm{T}}, \theta^{\mathrm{T}})$. Suppose that the length of realization is n. Usually F_t will be the reduced information set $\{X_t, \ldots, X_{t-q}, y_{t-1}, \ldots, y_{t-q}\}$. Using the quasi-likelihood approach the estimating equation for γ conditional on the first q observations is

$$U(\gamma) = \sum_{t=q+1}^{n} Z_t \frac{\partial \mu_t}{\partial g_t}(y_t - \mu_t)/V_t = 0 , \tag{8.7}$$

where

$$Z_t^{\mathrm{T}} = \{X_t^{\mathrm{T}}, f_1(F_t), \ldots, f_q(F_t)\}, \quad g_t = g(\mu_t), \quad V_t = V(\mu_t) .$$

With canonical links the left-hand side of (8.7) simplifies to $\sum Z_t(y_t - \mu_t)$. In many applications $f_i(F_t) = f_i(F_{t-i})$. For example, for binary outcomes we may have

$$\text{logit}\ (\mu_t) = X_t^{\mathrm{T}}\beta + \theta_1 y_{t-1} + \cdots + \theta_q y_{t-q}\ .$$

As in the i.i.d. case iteratively reweighted least squares can be used to solve (8.7).

Under regularity conditions (Kaufmann, 1987; Fahrmeir and Kaufmann, 1987) it can be shown that $\sqrt{n}(\hat{\gamma} - \gamma)$ is asymptotically normally distributed with variance $\phi \boldsymbol{V}^{-1}$, where

$$\boldsymbol{V} = \lim_{n\to\infty} \frac{1}{n} \sum_{t=q+1}^{n} Z_t(\partial\mu_t/\partial g_t)^2 V_t^{-1} Z_t^{\mathrm{T}}\ .$$

For canonical links $\boldsymbol{V}$ simplifies to $\lim n^{-1} \sum Z_t V_t Z_t^{\mathrm{T}}$ (Zeger and Qaqish, 1988). Note that the value of ϕ does not affect the estimation of γ and can be estimated as

$$\hat{\phi} = \frac{1}{n} \sum_{t=q+1}^{n} \hat{a}_t^2\ ,$$

where $\hat{a}_t = (y_t - \hat{\mu}_t)/\boldsymbol{V}(\hat{\mu}_t)^{\frac{1}{2}}$. Li (1991) derived the asymptotic distribution for the autocorrelation of $\hat{a}_t$.

To obtain Li's result we define a_t and r_k slightly differently. Let $a_t = (y_t - \mu_t)/V(\mu_t)^{\frac{1}{2}}$. Then the lag k innovation autocorrelation r_k is given by

$$r_k = \frac{1}{n} \sum_{t=k+1}^{n} a_t a_{t-k}/\phi \quad (k = 1, \ldots, m)\ .$$

Let $\boldsymbol{r} = (r_1, \ldots, r_m)^{\mathrm{T}}$ for some $m > 0$. Similarly define the residual autocorrelations $\hat{r}_k$ by

$$\hat{r}_k = \frac{1}{n} \sum \hat{a}_t \hat{a}_{t-k}/\hat{\phi} \quad (k = 1, \ldots, m)\ ,$$

Let $\hat{\boldsymbol{r}} = (\hat{\boldsymbol{r}}_1, \ldots, \hat{\boldsymbol{r}}_m)^{\mathrm{T}}$. Note that $\{a_t\}$ is a sequence of martingale differences with finite variance.

As in Chapter 2 the following theorem in Li (1991) can be proved by the method of McLeod (1978).

Theorem 8.1 *If the model is correct, $\hat{\boldsymbol{r}}\sqrt{n}$ is asymptotically normally* *distributed with mean zero and variance* $\mathbf{1}_m - \phi^{-1}\boldsymbol{X}^{\mathrm{T}}\boldsymbol{V}^{-1}\boldsymbol{X}$, *where*

$$\boldsymbol{X} = \lim_{n\to\infty} n^{-1} \begin{bmatrix} \sum X_t h_t a_{t-1} & \sum X_t h_t a_{t-2} & \cdots & \sum X_t h_t a_{t-m} \\ \sum f_1(t) h_t a_{t-1} & \sum f_1(t) h_t a_{t-2} & \cdots & \sum f_1(t) h_t a_{t-m} \\ \vdots & \vdots & & \vdots \\ \sum f_q(t) h_t a_{t-1} & \sum f_q(t) h_t a_{t-2} & \cdots & \sum f_q(t) h_t a_{t-m} \end{bmatrix},$$

where we write $f_i(t) = f_i(F_t)$ *and* $h_t = V_t^{-\frac{1}{2}} \partial\mu_t/\partial g_t$.

Note that if $f_i(F_t) = f(F_{t-i})$ and h_t is a constant, then $n^{-1}\sum f_i(t)h_t a_{t-j}$ converges to zero if $i > j$. Further simplification results if X_t and a_{t-i} are uncorrelated. If y_t has the usual autoregressive moving average structure with $V_t = 1$ and $\phi = \sigma^2$, then we obtain the usual portmanteau statistic. In many applications, $(\hat{\boldsymbol{r}}^{\mathrm{T}}, \hat{\phi})$ and the sample averages $\sum f_i(t)\hat{h}(t)\hat{a}_{t-j}/n$ can be substituted into $\mathbf{1}_m - \phi^{-1}\boldsymbol{X}^{\mathrm{T}}\boldsymbol{V}^{-1}\boldsymbol{X}$ to obtain the standard errors for $\hat{r}_i$. An overall test for the significance of residual autocorrelations can also be based on $n\hat{\boldsymbol{r}}^{\mathrm{T}}(\mathbf{1}_m - \hat{\phi}^{-1}\hat{\boldsymbol{X}}^{\mathrm{T}}\hat{\boldsymbol{V}}^{-1}\hat{\boldsymbol{X}})^{-1}\hat{\boldsymbol{r}}$, which is asymptotically chi-squared with m degrees of freedom.

We now derive a score test for testing for a possible higher order model as in Li (1991). Let $\gamma_1^{\mathrm{T}} = (\beta^{\mathrm{T}}, \theta_1, \ldots, \theta_{q_0})$, $\gamma_2^{\mathrm{T}} = (\theta_{q_0+1}, \ldots, \theta_{q_0+k})$, $\gamma^{\mathrm{T}} = (\gamma_1^{\mathrm{T}}, \gamma_2^{\mathrm{T}})$. The null hypothesis is $\gamma_2^{\mathrm{T}} = 0$ against the alternative that $q = q_0 + k$. The corresponding score statistic is simply $U(\gamma) = \sum Z_t(\partial\mu_t/\partial g_t)(y_t - \mu_t)/V_t$, where $Z_t^{\mathrm{T}} = (Z_{1t}^{\mathrm{T}}, Z_{2t}^{\mathrm{T}})$ with

$$Z_{1t}^{\mathrm{T}} = (X_t^{\mathrm{T}}, f_1(F_t), \ldots, f_{q_0}(F_t)), \quad Z_{2t}^{\mathrm{T}} = (f_{q_0+1}(F_t), \ldots, f_{q_0+k}(F_t)) .$$

It can be shown that $U(\gamma)/\sqrt{n}$ is asymptotically normally distributed with mean zero and variance $\phi\boldsymbol{V}$. Let $\boldsymbol{V}$ be partitioned according to $\gamma^{\mathrm{T}} = (\gamma_1^{\mathrm{T}}, \gamma_2^{\mathrm{T}})$. Denote this partition as $\boldsymbol{V} = (V_{ij})$, for $i, j = 1, 2$. Following Basawa (1985) and Serfling (1980, Ch.4) a score or Lagrange multiplier statistic for testing the above hypotheses is given by

$$\begin{aligned} \mathrm{LM} &= n^{-1}\hat{U}(\gamma_2)^{\mathrm{T}}(\hat{V}_{22} - \hat{V}_{21}\hat{V}_{11}^{-1}\hat{V}_{12})^{-1}\hat{U}(\gamma_2)/\hat{\phi} \\ &= n^{-1}\sum Z_{2t}^{\mathrm{T}}\frac{\partial\mu_t}{\partial g_t}V_t^{-1}(y_t - \hat{\mu}_t)(\hat{V}_{22} - \hat{V}_{21}\hat{V}_{11}^{-1}\hat{V}_{12})^{-1} \\ &\quad \cdot \sum Z_{2t}\frac{\partial\mu_t}{\partial g_t}V_t^{-1}(y_t - \hat{\mu}_t)/\hat{\phi} , \end{aligned} \tag{8.8}$$

where $\hat{V}_{ij}$, $\hat{\phi}$, and $\hat{\mu}_t$ are evaluated under the null model. Under the null model, LM is asymptotically chi-squared with k degrees of freedom. Evaluation of (8.8) may seem complicated. However, we may rewrite

(8.8) as

$$\mathrm{LM} = (n\hat{\phi})^{-1}\left[\sum (y_t - \hat{\mu}_t)\hat{V}_t^{-1}\frac{\partial \mu_t}{\partial g_t} Z_t^{\mathrm{T}}\right]\hat{\boldsymbol{V}}^{-1}\left[\sum Z_t \frac{\partial \mu_t}{\partial g_t}\hat{V}_t^{-1}(y_t - \hat{\mu}_t)\right], \tag{8.9}$$

noting that $U(\hat{\gamma}_1) = 0$ under the null hypothesis. Let

$$\hat{a}_t = \frac{y_t - \hat{\mu}_t}{\hat{V}_t^{\frac{1}{2}}}, \qquad W_t = \left(\frac{\partial \mu_t}{\partial g_t}\right)\hat{V}_t^{-\frac{1}{2}} Z_t = h_t Z_t ,$$

$$a^{\mathrm{T}} = (\hat{a}_1, \ldots, \hat{a}_n), \quad W^{\mathrm{T}} = (W_1, \ldots, W_n) .$$

Then (8.9) can be rewritten as

$$\mathrm{LM}_* = a^{\mathrm{T}} W \Big(\lim_{n\to\infty} n^{-1} W^{\mathrm{T}} W\Big)^{-1} W^{\mathrm{T}} a/\hat{\phi} n .$$

Define, for n large enough,

$$\mathrm{LM}' = a^{\mathrm{T}} W (W^{\mathrm{T}} W)^{-1} W^{\mathrm{T}} a/\hat{\phi} .$$

For large samples LM′ and LM will have the same asymptotic distribution. Note that for canonical links $W_t = \hat{V}_t^{\frac{1}{2}} Z_t$. Furthermore we note as in earlier chapters that LM′ is n times the coefficient of determination, R^2, of the usual ordinary regression of a on W. Recall that $\hat{\phi} = a^{\mathrm{T}} a/n$. Consequently a test of $q = q_0$ against $q = q_0 + k$ can be based on nR^2 of a one-step auxiliary regression. Note that Pregibon (1982) has proposed a score statistic in the context of generalized linear models. His statistic is also based on a similar auxiliary regression but the interpretation is different in that his score statistic is the difference between two Pearson chi-squared statistics rather than the nR^2 here.

Example 8.1 *Neuron impulse data (Li, 1991). ©1991 Biometrika Trust, reproduced with the permission of Oxford University Press*

We considered the neuron impulse data of Zeger and Qaqish (1988). Two models were given by these authors. In the first model we have

$$\frac{1}{\mu_t} = \mu + \sum_{i=1}^{2} \theta_i \left(\frac{1}{y_{t-i}} - \mu\right), \tag{8.10}$$

with $var(y_t) = \mu_t^2 \phi$. *The time series was assumed to be conditionally distributed as Gamma with a constant coefficient of variation. The second model was given by adding the spike sequence number to (8.10) as a trend variable. Two score tests* LM_1' *and* LM_3' *were considered. The statistic* LM_1' *tested the null hypothesis* $q = 2$ *vs. the alternative* $q = 3$, *and* LM_3' *tested the null hypothesis* $q = 2$ *vs. the alternative* $q = 5$.

Using the estimates of Zeger and Qaqish (1988) as initial values in the estimation of (8.10) we have

$$(\hat{\mu}, \hat{\phi}, \hat{\theta}_1, \hat{\theta}_2) = (0.0249, 0.2975, 0.0953, 0.1160) .$$

The values of LM_1' *and* LM_3' *were* 4.673 *and* 11.745*, respectively. The corresponding 5% critical values for* LM_1' *and* LM_3' *are given by* 3.841 *and* 7.815 *indicating that the model was not adequate. This finding is different from that of Zeger and Qaqish (1988). The model (8.10) was considered adequate based on the bootstrap distribution of the residual autocorrelations and the deviance. When a trend was included,* $(\hat{\mu}, \hat{\phi}, \hat{\theta}_1, \hat{\theta}_2) = (0.0133, 0.2114, 0.0326, 0.0426)$ *and the coefficient for the trend was found to be* 0.000297*. The values of* LM_1' *and* LM_3' *were* 2.465 *and* 5.671 *which were not significant at the respective 10% levels. Hence, although model (8.10) was rejected by the score statistics, the final result did suggest that the trend model was justified. Note that our estimates of* θ_1 *and* θ_2 *were somewhat smaller than Zeger and Qaqish's. From these results it seems that LM′ can be a useful diagnostic tool when used with care.*

Li (1994) considered the possibility of introducing moving average terms to (8.6) by enlarging F_t to include $\mu_{t-1}, \ldots, \mu_{t-k}$ for some $k < n$. Thus a more general formulation of (8.6) would be

$$\eta(\mu_t) = \sum_{i=1}^{r} \alpha_i g_i(F_t) , \tag{8.11}$$

where $F_t = \{X_t, \ldots, X_{t-k}, y_{t-1}, \ldots, y_{t-k}, \mu_{t-1}, \ldots, \mu_{t-k}\}$, $k < n$, and g_i are known functions. Let $\alpha^{\mathrm{T}} = (\alpha_1, \ldots, \alpha_r)$. This formulation is rather general and allows us a lot of flexibility. For example, in (8.6), we can consider $\ln y_{t-i}^* - \ln \mu_{t-i} = \ln y_{t-i}^*/\mu_{t-i}$. As the simulation in Li (1994) shows, the time series does give an autocorrelation structure that is typical of the classical moving average models. In any case, we may regard (8.11) as a generalized autoregressive moving average model. Further extension of this idea has been taken up recently by Benjamin, Ribby, and Stasinopoulos (2003).

Suppose y_t is an invertible time series. Let g_i be differentiable functions of μ_{t-j}, $j = 1, \ldots, k$. Estimation of (8.11) can be based on the quasi-likelihood approach. However, μ_t now depends, in an iterative sense, on all previous observations. We may compute μ_t by setting initial μ_t's to zero or to the sample mean of y_t. Likewise the derivatives of (8.11) with respect to α_i will also involve all previous observations. Consider

$$\frac{\partial \eta_t}{\partial \alpha_j} = g_i(F_t) + \sum_{i=1}^{r} \alpha_i \frac{\partial g_i(F_t)}{\partial \alpha_j} , \quad j = 1, \ldots, r .$$

where $\eta_t = \eta(\mu_t)$. Let $g_i(t) = g_i(F_t)$ then,

$$\frac{\partial g_i(t)}{\partial \alpha_j} = \sum_{l=1}^{k} \frac{\partial g_i(t)}{\partial \mu_{t-l}} \frac{\partial \mu_{t-l}}{\partial \alpha_j} = \sum_{l=1}^{k} \left(\frac{\partial g_i(t)}{\partial \mu_{t-l}} \right) \frac{\partial \eta_{t-l}}{\eta'_{t-l} \partial \alpha_j} ,$$

where $\eta'_t = \partial \eta_r / \partial \mu_t$ and for the canonical link, $\eta'_t = V_t^{-1}$. Now $\partial \eta_t / \partial \alpha_j$ can be computed recursively by setting

$$\frac{\partial \eta_0}{\partial \alpha_j} = \cdots = \frac{\partial \eta_{1-k}}{\partial \alpha_j} = 0 , \quad j = 1, \ldots, r .$$

Denote estimates of $\boldsymbol{\alpha}$ by $\hat{\boldsymbol{\alpha}}$. The quasi-likelihood estimating equations are then

$$\sum_{t=1}^{n} Z_t \cdot \frac{y_t - \mu_t}{\eta'_t V_t} = 0 ,$$

where $Z_t = \partial \eta_t / \partial \boldsymbol{\alpha}$. Starting with an initial value $\boldsymbol{\alpha}_0$ sufficiently close to $\hat{\boldsymbol{\alpha}}$, the estimates can be obtained iteratively as in McCullagh and Nelder (1989, p.327) by Fisher scoring. Similar to Li (1991) we can derive LM tests for testing model adequacy.

Example 8.2 *The U.S. Poliomyelitis data (Li, 1994).* *©1994 International Biometric Society, reproduced with the permission of Blackwell Publishing*

As an example we consider the U.S. poliomyelitis data (1970–1983) in Zeger (1988). It is of interest to know whether there is a long-term decrease in the U.S. polio infection rate. Zeger considered a parameter-driven model and found that if a first-order autoregression was assumed for the latent process, the evidence for a decreasing trend became much weaker. However, he also found significant first-order residual autocorrelation in his model. This suggested that some higher-order latent processes may be needed to take care of the autocorrelation structure. Estimation, however, would then be more difficult with the parameter-driven approach. A more natural approach is to consider simply the observation-driven models. In Li (1994) four observation-driven Poisson models were considered. The first two are second-order autoregressive models with link functions similar to Zeger and Qaqish (1988, eq.2.2), namely,

$$\ln(\mu_t) = \mu + \beta t + \phi_1 \ln y_{t-1} + \phi_2 \ln y_{t-2} ,$$

where t is the case number and $\beta = 0$ for the first model. To avoid zeros while at the same time preserving autocorrelation structure, we have added a value of .1 to all data. Models 3 and 4 are second-order moving average models with link functions

$$\ln(\mu_t) = \mu + \beta t + \theta_1 \ln y_{t-1}/\mu_{t-1} + \theta_2 \ln y_{t-2}/\mu_{t-2} .$$

Again, Model 3 assumes $\beta = 0$. *The estimation results are reported in Table 8.2 together with the deviance (Dev), the residual mean square (RSS), and the values of a score statistic (LM) for testing whether ten more lags are needed in the respective autoregressive and moving average models. It can be seen that, based on the deviance, the best model is Model 4, the second-order moving average with trend. Judging from the score statistics, the two autoregressive models do not seem to be able to capture the autocorrelation structure adequately. The deviance and the residual mean square of the autoregressive models are also higher than those of the moving average models. For Model 4, the likelihood ratio test for trend based on the difference in deviance is not significant, although the sign of* $\hat{\boldsymbol{\beta}}$ *is negative as expected and its value is twice that of Model 2. Observe that* $\hat{\boldsymbol{\beta}}$ *in Model 2 is significant at the 10% level. However, from the score statistic, Model 2 appears to have some significant residual autocorrelations. This is in some way similar to Zeger's (1988) result where the lag one residual autocorrelation was also significant. Here, the significance of the trend estimate was further reduced by the moving average models. It may appear controversial that the evidence for a decreasing trend in polio infection is almost nonexistent after accounting for the autocorrelation structure. A visual display of the data suggests that the total number of infectious cases in later years may not be too different from some of the earlier ones. Thus the present sample size may be too small to give a significant* $\hat{\boldsymbol{\beta}}$. *In any case, statistical inference should be more valid, when residual autocorrelations have been fully accounted for as in the moving average models.*

Table 8.2 *Estimation results for the U.S. poliomyelitis data. From Li (1994).*

Model	$\hat{\mu}$	$\hat{\beta}$	$\hat{\phi}_1$	$\hat{\phi}_2$	Dev.	RSS	LM
1	.729	—	.224	.127	261.26	3.15	17.26
2	.990	−.0025	.211	.114	257.94	3.11	17.11
Model	$\hat{\mu}$	$\hat{\beta}$	$\hat{\theta}_1$	$\hat{\theta}_2$	**Dev.**	**SS**	**LM**
3	.605	—	.260	.232	250.09	2.96	9.65
4	1.004	−.0053	.243	.221	247.91	2.93	10.31

These results suggest that the proposed method of defining moving average models and the corresponding modeling procedures can be of potential practical use. Note that statistical inferences are much easier using the current conditional distribution approach than the marginal distribution approach.

8.3 The autoregressive conditional duration model

The autoregressive conditional duration (ACD) model proposed by Engle and Russell (1997, 1998) is a new statistical model for analyzing a sequence of time events which arrive at irregular intervals and have possibly high intertemporal correlation. A typical example is the stock transaction duration data collected in financial markets. Figure 8.2 shows the transaction duration of the Hong Kong stock, Cheung Kong Holdings(0001), on December 1, 1988. The data are available from the Hong Kong Exchanges and Clearing Ltd. We can see that the transaction durations are fairly short during the first 20 minutes of trading, while they are quite long at the time around 15 minutes to 11:00 AM. Durations are generally longer toward the middle of the day in the morning session. Transaction durations are much longer at the opening of the afternoon session and then they are extremely short during the 10 to 15 minutes before the market closing at 4:00 AM. This clustering of transactions can be further evidenced by a high autocorrelation between successive transaction durations.

Because of the above-mentioned special structure of transaction duration data, standard time series techniques are not applicable as they deal mainly with data recorded in regular time intervals. One way to employ these methods is to aggregate the irregular transaction to the one measured in a regular time interval basis such as daily or weekly basis. This however causes problems. Many zero information observations will be created if a short time interval is chosen. On the other hand, finer structure information will be lost if a long time interval is chosen. This problem becomes much worse when the data contain intra-day patterns. In fact, the transaction duration data can be regarded as a kind of survival or lifetime data. More specifically, the time for a new transaction to occur can be more or less treated as the survival time of a patient after a medical treatment, or more generally the failure time until an event occurs. In the literature, many well-known statistical models have been proposed for lifetime data. However, these models cannot be directly applied to model transaction duration data. The main reason is that in survival analysis, the individuals under study and hence their lifetimes are independent while, as pointed out before, transaction duration data are highly autocorrelated. As a consequence, a new modeling techniques for intertemporally correlated irregular duration data have recently been developed. In this paper, we will focus on a new class of models called

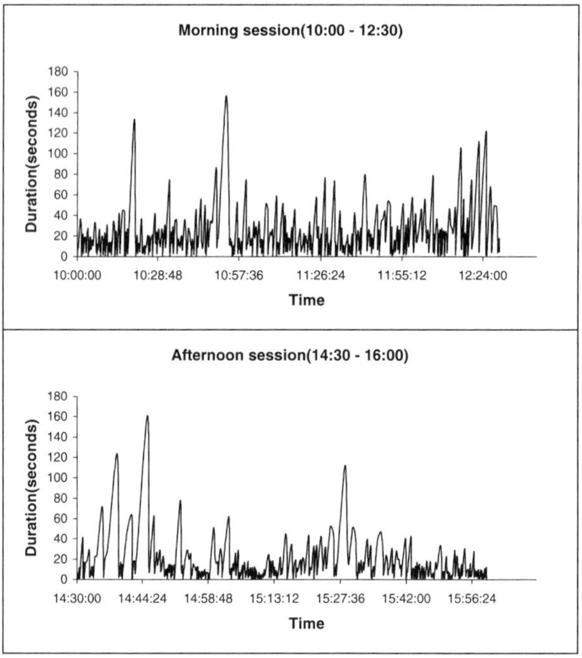

Figure 8.2 *Transaction duration of a stock throughout a whole trading day*

Autoregressive Conditional Duration (ACD) models, proposed by Engle and Russell (1997, 1998), which can help to explain such phenomena.

The ACD model has since become very popular in the modeling of time series of duration data, especially in finance. Following Engle and Russell (1997, 1998), numerous other models with features of ACD have been proposed. A diagnostic test based on the residual autocorrelations for the ACD model has been developed in Li and Yu (2003).

Let x_t be the duration process of interest. Let $\mathcal{F}_t$ be the information set generated by all past observations up to and including the t-th transaction. The exponential ACD model for x_t is defined as,

$$x_t = \psi_t e_t \quad \psi_t = \omega + \sum_{j=1}^{p} \alpha_j x_{i-j} \ , \tag{8.12}$$

where $\omega > 0$, $\alpha_j \geq 0$. Here we treat t as if it were chronological time.

We assume e_t to follow the standard exponential distribution. The general case with e_t following a Weibull $(1, \gamma)$ distribution can be easily handled by the transformation x^γ. Note that $E(x_t|\mathcal{F}_{t-1}) = \psi_t$. For stability of (8.12) it is assumed, as in the autoregressive conditional heteroscedastic (ARCH) case, that $\sum_{j=1}^{p} \alpha_j < 1$.

Let $\boldsymbol{\theta}$ be the vector of parameters $(\omega, \alpha_1, \ldots, \alpha_p)^{\mathrm{T}}$ and $\hat{\boldsymbol{\theta}}$ be the conditional maximum likelihood estimator of $\boldsymbol{\theta}$. Let $\hat{e}_t$ be the corresponding residual when $\boldsymbol{\theta}$ is replaced by $\hat{\boldsymbol{\theta}}$. The lag-$k$ residual autocorrelation is defined as

$$\hat{r}_k = \frac{\sum\limits_{i=k+1}^{n} (\hat{e}_t - \bar{e})(\hat{e}_{t-k} - \bar{e})}{\sum\limits_{i=1}^{n} (\hat{e}_t - \bar{e})^2} , \quad k = 1, 2, \ldots, m; \; t = 1, \ldots, n .$$

Denote the corresponding lag-k sample autocorrelation of e_t by r_k. Since it can be shown that $\bar{e} = \sum \hat{e}_t / n \xrightarrow{p} 1$ if the model is correct, and $\sum(\hat{e}_t - 1)^2/n$ converges also to 1 in probability, we need only consider the asymptotic distribution of $\hat{\boldsymbol{r}} = (\hat{r}_1, \hat{r}_2, \ldots, \hat{r}_M)^{\mathrm{T}}$ where

$$\hat{r}_k = \frac{\sum\limits_{i=k+1}^{n} (\hat{e}_t - 1)(\hat{e}_{t-k} - 1)}{n} .$$

As before $\sqrt{n}\boldsymbol{r} = \sqrt{n}(r_1, \ldots, r_m)^{\mathrm{T}}$ is asymptotically $N(0, \mathbf{1}_m)$ distributed, where $\mathbf{1}_m$ is the $m \times m$ identity matrix.

First, following Li and Yu (2003), we examine the asymptotic distribution and the information matrix $\mathbf{I}$ of $\hat{\boldsymbol{\theta}} = (\hat{\omega}, \hat{\alpha})^{\mathrm{T}}$. Let $x_0 = 0$ and $\psi_0 = 1$. For each t, we denote the conditional log-likelihood of x_t by ℓ_t and $\ell_t = -\log \psi_t - x_t/\psi_t$. Then the conditional log-likelihood of the data is given by $\ell = \sum_{t=1}^{n} \ell_t$.

For ease of exposition and without loss of generality, let $p = 1$ and $\alpha_1 = \alpha$. By direct differentiation of the log-likelihood using the result $\partial\psi_t/\partial\psi = 1$, and $\partial\psi_t/\partial\alpha = x_{t-1}$, we have,

$$\begin{aligned}
\frac{\partial L}{\partial \omega} &= -\sum_{t=1}^{n} \left[\left(\frac{1}{\psi_t} \right) \frac{\partial \psi_t}{\partial \omega} - \left(\frac{x_t}{\psi_t^2} \right) \frac{\partial \psi_t}{\partial \omega} \right] \\
&= -\sum_{t=1}^{n} \frac{1}{\psi_t} (1 - e_t) , \\
\frac{\partial L}{\partial \alpha} &= -\sum_{t=1}^{n} \left[\frac{1}{\psi_t} \frac{\partial \psi_t}{\partial \alpha} - \frac{x_t}{\psi_t^2} \frac{\partial \psi_t}{\partial \alpha} \right] \\
&= -\sum_{t=1}^{n} \left(\frac{x_{t-1}}{\psi_t} \right) (1 - e_t) .
\end{aligned}$$

Differentiating again we have

$$\begin{aligned}
\frac{\partial^2 L}{\partial \omega^2} &= \sum_{t=1}^{n} \frac{\partial}{\partial \omega}\left[-\frac{1}{\psi_t}+\frac{x_t}{\psi_t^2}\right] \\
&= \sum_{t=1}^{n}\left(\frac{1}{\psi_t^2}-2\frac{x_t}{\psi_t^3}\right), \\
\frac{\partial^2 L}{\partial \alpha \partial \omega} &= \sum_{t=1}^{n}\left[-\frac{2x_t x_{t-1}}{\psi_t^3}+\frac{x_{t-1}}{\partial_t^2}\right], \\
\frac{\partial^2 L}{\partial \alpha^2} &= \sum_{t=1}^{n}\left(\frac{-2x_t x_{t-1}^2}{\psi_t^3}+\frac{x_{t-1}^2}{\psi_t^2}\right).
\end{aligned}$$

Under the usual regularity conditions, $\sqrt{n}(\hat{\boldsymbol{\theta}}-\boldsymbol{\theta})$ can be shown to be asymptotically normal with zero mean and covariance matrix $\mathbf{I}^{-1} = -E(n^{-1}\partial^2 L/\partial\boldsymbol{\theta}\partial\boldsymbol{\theta}^{\mathrm{T}})^{-1}$. Now we turn to the asymptotic distribution of $\hat{\boldsymbol{r}}$. As in Li and Yu (2003), using Taylor series expansion, $\hat{\boldsymbol{r}}$ can be expressed asymptotically as

$$\hat{\boldsymbol{r}} \sim \boldsymbol{r} - \boldsymbol{X}(\hat{\boldsymbol{\theta}}-\boldsymbol{\theta}) ,$$

where X is an $M \times 2$ matrix

$$\boldsymbol{X} = \begin{pmatrix} \frac{1}{n}\sum_{t=2}^{n}\frac{x_t(e_{t-1}-1)}{\psi_t^2} & \frac{1}{n}\sum_{t=2}^{n}\frac{x_t x_{t-1}(e_{t-1}-1)}{\psi_t^2} \\ \vdots & \vdots \\ \frac{1}{n}\sum_{t=M+1}^{n}\frac{x_t(e_{t-M}-1)}{\psi_t^2} & \frac{1}{n}\sum_{t=M+1}^{n}\frac{x_t x_{t-1}(e_{t-M}-1)}{\psi_t^2} \end{pmatrix}.$$

As in chapter 2 the vector $\hat{\boldsymbol{r}}$ can be shown to be asymptotically normally distributed by the martingale central limit theorem.

Theorem 8.2 *(Li and Yu, 2003)*

The large sample distribution of $\sqrt{n}\hat{\boldsymbol{r}}$ is normal with mean $\mathbf{0}$ and covariance matrix $\mathbf{1} - \boldsymbol{X}\mathbf{I}^{-1}\boldsymbol{X}^{\mathrm{T}}$. Here $\mathbf{I} = -E\big(n^{-1}\partial^2\ell/\partial\boldsymbol{\theta}\partial\boldsymbol{\theta}^{\mathrm{T}}\big)$.

In practice, we can estimate the entries of G by its sample average. The statistic $Q = n\hat{\boldsymbol{r}}^{\mathrm{T}}(\mathbf{1} - \boldsymbol{X}\mathbf{I}^{-1}\boldsymbol{X}^{\mathrm{T}})^{-1}\hat{\boldsymbol{r}}$ will be asymptotically χ^2 distributed with M degrees of freedom if the fitted model is correct. For the general case $p > 1$, $\boldsymbol{X}$ is an $M \times (p+1)$ matrix with the kth row given by $\big(\frac{1}{n}\sum_{i=k+1}^{n}\psi_t^{-1}\partial\psi_t/\partial\boldsymbol{\theta}^{\mathrm{T}}(e_{t-k}-1)\big)$, $M \geq k \geq 1$. As in the case of ARMA models, more accurate asymptotic standard errors of $\hat{r}_k$ can be obtained from Theorem 8.2.

8.4 A power transformation to induce normality

Many statistical tests can be written as positive linear combinations of positive independent random variables. However, the finite sample distribution of these statistical tests could be highly skewed to the right although asymptotically they are normally distributed. Chen and Deo (2003) considered a power transformation which appears to improve the problem of skewness and hence improve the finite sample performance of such statistics. Let the transformation be $h(y) = y^{\beta}$. The idea is to obtain a β such that the skewness of y is approximately zero. Let the statistic be denoted by T_n.

Let $\{a_{j,n}\}$ be an array of positive real numbers such that $\sum_{j=1}^{n} a_{j,n}^{s} = O(p_n)$ for $s \geq 1$, where $p_n^{-1} + n^{-1}p_n \to 0$ as $n \to \infty$. Consider the variable

$$T_n = \sum_{j=1}^{n} a_{j,n} X_j \ . \tag{8.13}$$

where X_j are independent identically distributed random variables whose first three moments are known. Let $\mu = E(X_j)$ and $\sigma^2 = \text{var}(X_j)$. Consider the scaled variable T_n/p_n. Note that $E(T_n/p_n) = (\mu/p_n)\sum_{j=1}^{n} a_{j,n}$ and $\text{var}(T_n/p_n) = \sigma_Y^2 = (\sigma^2/p_n^2)\sum_{j=1}^{n} a_{j,n}^2$.

Using a Taylor expansion of $h(y) = y^{\beta}$ about the mean of T_n/p_n, Chen and Deo (2003) showed that the skewness of $h(T_n/p_n)$ is approximately zero if β is chosen to be

$$\beta = 1 - \frac{\mu E(X_1 - \mu)^3 (\sum_{i=1}^{n} a_{i,n})(\sum_{j=1}^{n} a_{j,n}^3)}{3\sigma^4 (\sum_{j=1}^{n} a_{j,n}^2)^2} \ . \tag{8.14}$$

They applied this transformation to Hong's test (see (6.23)),

$$H_n = n \cdot \sum_{j=1}^{p_n} k^2(j/p_n)\hat{\rho}(j)^2$$

and the generalized portmanteau test T_n of Chen and Deo (2001),

$$T_n = \left\{ \frac{2\pi}{n} \sum_{j=0}^{n-1} \hat{f}(\lambda_j) \right\}^{-2} \left\{ \frac{2\pi}{n} \sum_{j=0}^{n-1} \hat{f}^2(\lambda_j) \right\} , \tag{8.15}$$

where

$$\hat{f}(\lambda) = \frac{2\pi}{n} \sum_{j=1}^{n-1} \frac{W(\lambda - \lambda_j) I_x(\lambda_j)}{f(\lambda_j)} \ ,$$

$f(\cdot)$ is the spectral density of the fitted model and $I_x(\lambda) = (2\pi n)^{-1} |\sum_{t=1}^{n}$

$x_t \exp(-it\lambda)|^2$ is the periodogram of the observations x_t and

$$W(\lambda) = \frac{1}{2\pi} \sum_{|j|<n} k(j/p_n) e^{-ij\lambda} \qquad -\pi \leq \lambda \leq \pi \, .$$

According to their simulation results, the transformed tests H_n^β and T_n^β have much better performance than either $\tilde{Q}_m$ (2.12) or Q_m^* (2.13). However, without the transformation the performance of all four tests are comparable.

The β for Hong's test H_n is given by

$$1 - \frac{2}{3} \frac{(\sum_{j=1}^{p_n} k_j^2)(\sum_{j=1}^{p_n} k_j^6)}{(\sum_{j=1}^{p_n} k_j^4)^2} \, .$$

Of course, a major criticism of these tests is that there is no clear objective way to choose the bandwidth p_n and even if there is one such way it usually requires very heavy computation.

8.5 Epilogue

A challenge for the statistician is to choose the "best" model for his/her data. In many situations this goal can be partly fullfilled, at least, by applying various diagnostic tools to the fitted models, as is the case in time series. The portmanteau goodness-of-fit test is one of those simple and sensible tools for a time series analyst. We have seen, through the chapters, that it is possible to derive such a pure significance test in many situations. There have been many attempts to improve the power of such tests. This continues to be a challenging area. On the other hand, there are many other tests developed with specific alternatives in mind which are more powerful than the portmanteau test. This book has chosen to sample only a few of these due to space limitations. Extension of all these tests to more complicated models or situations will always be a challenge and adventure. As an example, developing goodness-of-fit tests in models that are fitted by the Markov chain Monte Carlo method appears to be an area that is under-researched. Another example is the mixture time series models of Wong and Li (2001). It is hoped that this book will help to attract more interest leading to fruitful research in the area of diagnostic checks in time series.

References

Abraham, B. and Chuang, A. (1989). Outlier detection and time series modelling. *Technometrics*, **31**, 241–248.

Abraham, B. and Ledolter, J. (1983). *Statistical Methods for Forecasting.* John Wiley, New York.

Agiakloglou, C. and Newbold, P. (1994). Lagrange multiplier tests for fractional difference. *J. Time Ser. Anal.*, **15**, 253–262.

Aguirre-Torres, V. and Gallant, R. (1982). The null and non-null asymptotic distribution of the Cox test for multivariate nonlinear regression alternatives and a new distribution free Cox test. *J. Economet.*, **21**, 5–33.

Ahn, S.K. (1988). Distribution for residual autocovariances in multivariate autoregressive models with structured parameterization. *Biometrika*, **75**, 590–593.

An, H.Z. and Cheng, B. (1991). A Kolmogorov-Smirnov type statistic with application to test for nonlinearity in time series. *International Statistical Review*, **59**, 287–307.

Anderson, T.W. (1958). *An Introduction to Multivariate Statistical Analysis.* Wiley, New York.

Ansley, C.F. (1979). An algorithm for the exact likelihood of a mixed autoregressive-moving average process. *Biometrika*, **66**, 59–65.

Ansley, C.F. (1980). Asymptotic distribution of residual autocorrelations from estimation of ARMA processes by Gram-Schmidt orthogonalization. *Stochastic Processes and Their Applications*, **11**, 201–206.

Ansley, C.F. (1980). Computation of the theoretical autocovariance function for a vector ARMA process. *J. Statist. Comput. Simul.*, **12**, 15–24.

Ansley, C.F. and Newbold, P. (1979). On the finite sample distribution of residual autocorrelations in autoregressive-moving average models. *Biometrika*, **66**, 547–553.

Atkinson, A. (1986). *Regression Diagnostics* in Encyclopedia of Statistical Science. Eds. S. Kotz and N.L. Johnson, **7**, pp. 689–694, Wiley, New York.

Baillie, R.T. (1996). Long memory processes and fractional integration in econometrics. *J. Economet.*, **73**, 5–59.

Bartlett, M.S. (1953). *Stochastic Processes.* Cambridge University Press, London.

Bartlett, M.S. (1966). *An Introduction to Stochastic Processes: with Special Reference to Methods and Applications.* Cambridge University Press, London.

Bartlett, M.S. and Diananda, P.H. (1950). Extensions of Quenouille's test for autoregressive schemes. *J. Roy. Statist. Soc.* B, **12**, 108–115.

Basawa, I.V. (1985). Neyman-le Cam tests based on estimating functions. In *Proceedings of the Berkeley Conference in Honor of Jerzy Neyman and Jack Kiefer*, **2**, Ed. L.M. Le Cam and R.A. Olshen, pp. 811–825, Wadsworth: Belmont.

Battaglia, F. (1990). Approximate power of portmanteau tests for time series. *Statistics & Probability Letters*, **9**, 337–341.

Benjamin, M.A., Ribby, R.A., and Stasinopoulos, D.M. (2003). Generalized autoregressive moving average models. *J. Amer. Statist. Ass.*, **98**, 214–223.

Beran, J. (1984). *Maximum Likelihood Estimation for Gaussian Processes with Long-range Dependence.* Abstract, 16th European Meeting of Statistician, Marburg, West Germany, p.71.

Beran, J. (1992). A goodness-of-fit test for time series with long range dependence. *J. Roy. Statist. Soc.* B, 749–760.

Beran, J. (1994). *Statistics for Long-memory Processes.* Chapman & Hall, New York.

Berks, I., Horváth, L., and Kokoszka, P. (2003). Asymptotics for GARCH squared residual autocorrelations. *Economet. Theor.*, **19**, 515–540.

Berndt, E.K., Hall, B.H., Hall, R.E., and Hausman, J.A. (1974). Estimation and inference in nonlinear structural models. *Annals of Economic and Social Measurement*, **4**, 653–665.

Billingsley, P. (1961). The Lindeberg-Levy theorem for martingales. *Proc. Am. Math. Soc.*, **12**, 788–792.

Billingsley, P. (1999). *Convergence of Probability Measures*, 2nd Edition. Wiley, New York.

Bollerslev, T. (1986). Generalized autoregressive conditional heteroskedasticity. *J. Economet.*, **31**, 307–327.

Bollerslev, T., Chou, R.Y., and Kroner, K.F. (1992). ARCH modeling in finance: a review of the theory and empirical evidence. *J. Economet.*, **52**, 5–59.

Bollerslev, T., Engle, R.F., and Nelson, D.B. (1994). ARCH models. In Handbook of Econometrics **4**, Ed. R.F. Engle and D.L. McFadden, North-Holland, Amsterdam.

Bollerslev, T., Engle, R.F., and Wooldridge, J.M. (1988). A capital asset pricing model with time varying covariance. *J. Polit. Econ.*, **96**, 116–131.

Bollerslev, T. and Wooldridge, J.M. (1992). Quasi-maximum likelihood estimation and inference in dynamic models with time-varying covariance. *Economet. Rev.*, **1**, 143–173.

Booth, G.G., Kaen, F.R., and Koveos, P.E. (1982). R/S analysis of foreign exchange rates under two international monetary regimes. *J. Monetary Economics*, **10**, 407–415.

Box, G.E.P. and Jenkins, G.M. (1970). *Time Series Analysis Forecasting and Control.* 1st edition. Holden-Day, San Francisco.

Box, G.E.P. and Jenkins, G.M. (1976). *Time Series Analysis Forecasting and Control.* 2nd edition. Holden-Day, San Francisco.

Box, G.E.P. and Pierce, D.A. (1970). Distribution of the residual autocorrelations in autoregressive integrated moving average time series models. *J. Amer. Statist. Ass.*, **65**, 1509–1526.

Box, G.E.P. and Tiao, G.C. (1975). Intervention analysis with applications to economic and environmental problems. *J. Amer. Statist. Ass.*, **70**, 70–79.

Box, G.E.P., Jenkins, G.M., and Reinsel, G.C. (1994). Time Series Analysis: *Forecasting and Control.* 3rd edition. Prentice-Hall, New Jersey.

Brillinger, D.R. (1965). An introduction to polyspectrum. *Ann. Math. Statist.*, **36**, 1351–1374.

Brockwell, P.J. and Davis, R.A. (1991). *Time Series: Theory and Methods.* 2nd Ed. Springer-Verlag, New York.

Brockwell, P.J. and Davis, R.A. (1996). *Introduction to Time Series and Forecasting.* Springer, New York.

Bruce, A.G. and Martin, R.D. (1989). Leave-k-out diagnostic for time series. *J. Roy. Statist. Soc.* B, **3**, 363–424.

Bustos, O.H., Fraiman, R., and Yohai, V.J. (1984). Asymptotic behaviour of the estimates based on residual autocovariances for ARMA models. In *Robust and Nonlinear Time Series Analysis*, Ed. J. Franke, W. Härdle, and D. Martin, pp.26–49. Springer-Verlag, New York.

Bustos, O.H. and Yohai, V.J. (1986). Robust estimates for ARMA models. *J. Amer. Statist. Ass.*, **81**, 155–168.

Chan, K.S. (1991). Percentage points of likelihood ratio tests for threshold autoregression. *J. Roy. Statist. Soc.* B, **53**, 691–696.

Chan, K.S. (1993). Consistency and limiting distribution of the least squares estiamtion of a threshold autoregressive model. *Ann. Statist.*, **21**, 520–533.

Chan, K.S., Petruccelli, J.D., Tong, H., and Woolford, S.W. (1985). A multiple threshold AR(1) model. *J. Appl. Prob.*, **22**, 267–279.

Chan, K.S. and Tong, H. (1986). On estimating threshold in autoregressive models. *J. Time Ser. Anal.*, **7**, 179–190.

Chan, K.S. and Tong, H. (1990). On likelihood ratio tests for threshold autoregression. *J. Roy. Statist. Soc.* B, **52**, 469–476.

Chan, W.S. (1994). On portmanteau goodness-of-fit test in robust time series modelling. *Comput. Statist.*, **9**, 301–310.

Chan, W.S. and Wei, W.W.S. (1992). A comparison of some estimators of time series autocorrelation. *Comput. Statist. Data Anal.*, **14**, 149–163.

Chan, W.Y.T. and Wallis, K.F. (1978). Multiple time series modelling: Another look at the mink-muskrat interaction. *Ann. Statist.*, **27**, 168–175.

Chang, I., Tiao, G.C., and Chen, C. (1988). Estimation of time series parameters in the presence of outliers. *Technometrics*, **30**, 193–204.

Chatfield, C. (1976). Comments on "Modelling macroeconomic time series" by D.L. Prothero and K.F. Wallis. *J. Roy. Statist. Soc.* A, **139**, 491.

Chen, W.W. and Deo, R.S. (2001). A generalized portmanteau goodness-of-fit test for time series models. Working Paper, New York University.

Chen, W.W. and Deo, R.S. (2003). Power transformation to induce normality and their applications. To appear in *J. Roy. Statist. Soc.* B.

Chernick, M.R., Downing, D.J., and Pike, D.H. (1982). Detecting outliers in time series data. *J. Amer. Statist. Ass.*, **77**, 743–747.

Cheung, Y.W. (1993). Long memory in foreign-exchange rates. *J. Bus. Econ. Stat.*, **11**, 93–101.

Cheung, Y.W. and Lai, K.S. (1993). A fractional cointegration analysis of purchasing power parity. *J. Bus. Econ. Stat.*, **11**, 103–112.

Cheung, Y.W. and Ng, L.K. (1996). A causality in variance test and its applications to financial market prices. *J. Economet.*, **72**, 33–48.

Chitturi, R.V. (1974). Distribution of residual autocorrelations in multiple autoregressive schemes. *J. Amer. Statist. Ass.*, **69**, 928–934.

Chong, T.T.L. (2000). Estimating the differencing parameter via the partial autocorrelation function. *J. Economet.*, **97**, 365–381.

Cox, D.R. (1962). Further results on tests of separate families of hypotheses. *J. Roy. Statist. Soc.* B, **24**, 406–424.

Cox, D.R. (1981). Statistical analysis of time series, some recent developments. *Scand. J. Statist.*, **8**, 93–115.

Cox, D.R. (2002). Karl Pearson and the chi-squared test: In *Goodness-of-fit Tests and Model Validity.* Eds. C. Huber-Carol, N. Balakrishnan, M.S. Nikulin, and M. Mesbah, pp.3–8, Birkhäuser: Boston.

Cryer, J.D. (1986). *Time Series Analysis.* Duxbury, Boston.

Dahlhaus, R. (1989). Efficient parameter estimation for self-similar processes. *Ann. Statist.*, **17**, 1749–1766.

Davies, N. and Petruccelli, J.D. (1986). Detecting non-linearity in time series. *The Statistician*, **35**, 271–280.

Davies, N., Spedding, T., and Watson, W. (1980). Autoregressive moving average processes with non-normal residuals. *J. Time Ser. Anal.*, **1**, 103–109.

Davies, N., Triggs, C.M., and Newbold, P. (1977). Significance levels of the Box-Pierce portmanteau statistic in finite samples. *Biometrika*, **64**, 517–522.

Davies, R.B. (1977). Hypothesis testing when a nuisance parameter is present only under the alternative. *Biometrika*, **64**, 247–254.

Davies, R.B. (1987). Hypothesis testing when a nuisance parameter is present only under the alternative. *Biometrika*, **74**, 33–43.

Demos, A. and Sentana, E. (1998). Testing for GARCH effects: a one sided approach. *J. Economet.*, **86**, 97–127.

Diebolt, J. (1990). Testing the functions defining a nonlinear autoregressive time series. *Stochastic Processes and Their Applications*, **36**, 85–106.

Ding, Z., Granger, C.W.J., and Engle, R.F. (1993). A long memory property of stock market returns and a new model. *J. Emprical Finance*, **1**, 83–106.

Duchesne, P. and Roy, R. (2003). Robust tests for independence of two time series. *Statistica Sinica*, **13**, 827–852.

Dufour, J.M. and Roy, R. (1985). Some robust exact results on sample autocorrelations and test of randomness. *J. Economet.*, **29**, 257–273.

Dufour, J.M. and Roy, R. (1986). Generalized portmanteau statistics and tests of randomness. *Commun. Statist.-Theor, Meth.*, **15**, 2953–2972.

Dunsmuir, W. and Robinson, P.M. (1981). Asymptotic theory for time series containing missing and amplitude modulated observations. *Sankhyā, Ser.* A, **43**, 260–281.

Durbin, J. (1970). Testing for serial correlation in least-squares regression when some of the regressors are lagged dependent variables. *Econometrica*, **38**, 410–421.

Efron, B. (1982). *The Jackknife, the Bootstrap and Other Resampling Plans.* Philadelphia: Society for Industrial and Applied Mathematics.

Efron, B. and Hinkley, D.V. (1978). Assessing the accuracy of the maximum likelihood estimator: observed versus expected Fisher information. *Biometrika*, **65**, 457–487.

Elton, C. and Nicholson, M. (1942). The ten-year cycle in numbers of the lynx in Canada. *J. Anim. Ecol.*, **11**, 215–244.

Engle, R.F. (1982). Autoregressive conditional heteroskedasticity with estimates of the variance of UK inflation. *Econometrica*, **50**, 987–1008.

Engle, R.F. (1983). Estimates of the variance of US inflation based on the ARCH model. *Journal of Money, Credit and Banking*, **15**, 286–301.

Engle, R.F. and Bollerslev, T. (1986). Modelling persistence of conditional variance. *Economet. Rev.*, **5**, 1–50.

Engle, R.F., Granger, C.W.J., and Kraft, D. (1984). Combining competing forecasts of inflation using a bivariate ARCH model. *J. Econ. Dyna. Cont.*, **8**, 151–165.

Engle, R.F. and Kroner, K.F. (1995). Multivariate simultaneous generalized ARCH. *Economet. Theor.*, **11**, 122–150.

Engle, R.F. and Russell, J.R. (1997). Forecasting the frequency of changes in quoted foreign exchange prices with the autoregressive conditional duration model. *J. Emprical Finance*, **4**, 187–212.

Engle, R.F. and Russell, J.R. (1998). Autoregressive conditional duration: a new model for irregularly-spaced transaction data. *Econometrica*, **66**, 1127–1162.

Fahrmeir, L. and Kaufmann, H. (1987). Regression models for nonstationary categorical time series. *J. Time Ser. Anal.*, **8**, 147–160.

Fisher, R.A. (1921). Contribution to the discussion of the paper, "On the time-correlation problem", by G.U. Yule, *J. Roy. Statist. Soc.*, **84**, 534–536.

Fox, A.J. (1972). Outliers in time series. *J. Roy. Statist. Soc.* B, **34**, 350–363.

Fox, R. and Taqqu, M.S. (1986). Large sample properties of parameter estimates for strongly dependent stationary Gaussian time series. *Ann. Statist.*, **14**, 517–532.

Fuller, W.A. (1996). *Introduction to Statistical Time Series.* 2nd edition. Wiley, New York.

Gabr, M.M. and Subba Rao, T. (1981). The estimation and prediction of subset bilinear time series models with applications. *J. Time Ser. Anal.*, **2**, 155–171.

Gaver, D.P. and Lewis, P.A.W. (1980). First-order autoregressive gamma sequences and point processes. *Adv. Appl. Prob.*, **12**, 727–745.

Geweke, J. and Porter-Hudak, S. (1983). The estimation and application of long memory time series models. *J. Time Ser. Anal.*, **4**, 221–238.

Ghaddar, D.K. and Tong, H. (1981). Data transformation and self-exciting threshold autoregression. *J. Roy. Statist. Soc.* C, **30**, 238–248.

Godfrey, L.G. (1978). Testing against general autoregressive and moving average error models when the regressors include lagged dependent variables. *Econometrica*, **46**, 1293–1301.

Godfrey, L.G. (1979). Testing the adequacy of a time series model. *Biometrika*, **66**, 67–72.

Godfrey, L.G. and Tremayne, A.R. (1988). Checks of model adequacy for univariate time series models and their applications to econometric relationships. EconometR, **7**, 1-42.

Godolphin, E.J. (1978). A large-sample test for detecting gaps in moving avarage models. *J. Roy. Statist. Soc.* B, **40**, 290–295.

Godolphin, E.J. (1980). A method for testing the order of an autoregressive-moving average models. *Biometrika*, **67**, 699-703.

Granger, C.W.J. (1969). Investigating causal relations by econometric models and cross-spectral methods. *Econometrica*, **37**, 424–438.

Granger, C.W.J. (1980a). Testing for causality: A personal viewpoint. *J. Econ. Dyna. Cont.*, **2**, 329–352.

Granger, C.W.J. (1980b). Long memory relationships and the aggregation of dynamic models. *J. Economet.*, **14**, 227–238.

Granger, C.W.J. and Andersen, A.P. (1978). *Introduction to Bilinear Time Series Models.* Vandenhoeck and Ruprecht, Göttingen.

Granger, C.W.J. and Joyeux, R. (1980). An introduction to long memory time series models and fractional differencing. *J. Time Ser. Anal.*, **1**, 15–39.

Granger, C.W.J. and Newbold, P. (1986). *Forecasting Economic Time Series.* 2nd edition. Academic Press, New York.

Granger, C.W.J., Spear, S., and Ding, Z. (2000). Stylized facts on the temporal and distributional properties of absolute returns: an update in *Statisitcs and Finance: An interface.* Eds. W.S. Chan, W.K. Li and H. Tong. Imperial College Press: London.

Granger, C.W.J. and Teräsvirta, T. (1993). *Modelling nonlinear economic relationships.* Oxford University Press, New York.

Grenander U. and Rosenblatt, M. (1957). *Statistical Analysis of Stationary Time Series.* Wiley, New York.

Guégan, D. (1981). Tests of adequacy for ARMA models and tests of separated hypothesis. In Stochastic Systems: *The Mathematics of Filtering and Identification and Applications*, (eds. M. Hazewinkel and J.C. Willems), 291–297, Dordrecht: Reidel.

Haggan, V. and Ozaki, T. (1981). Modelling non-linear random vibrations using an amplitude-dependent autoregressive time series model. *Biometrika*, **68**, 189–196.

Hall, P. and Heyde, C.C. (1980). *Martingale Limit Theory and Its Application.* Acadmeic Press, New York.

Hannan, E.J. (1969). The identification of vector mixed autoregressive-moving average systems. *Biometrika*, **56**, 223–225.

Hannan, E.J. (1970). *Multiple Time Series.* Wiley, New York.

Haugh, L.D. (1976). Checking the independence of two covariance-stationary time series: a univariate residual cross-correlation approach. *J. Amer. Statist. Ass.*, **71**, 378–385.

Haugh, L.D. and Box, G.E.P. (1977). Identification of dynamic regression (distributed lag) models connecting two time series. *J. Amer. Statist. Ass.*, **72**, 121–130.

Heathcote, C.R. and Welsh, A.H. (1988). Multivariate functional least squares. *J. Mult. Anal.*, **25**, 45–64.

Henrici, P. (1974). *Applied and computational complex analysis.* Vol. 1, Wiley, New York.

Higgins, M.L. and Bera, A.K. (1992). A class of nonlinear ARCH models. *International Economic Review*, **33**, 137–158.

Hill, B.M. (1963). The three-parameter lognormal distribution and Bayesian analysis of a point source epidemic. *J. Amer. Statist. Ass.*, **58**, 72–84.

Hipel, K.W. and McLeod, A.I. (1978). Preservation of the rescaled adjusted range. 2. Simulation studies using Box-Jenkins models. *Water Resour. Res.*, **14**, 509–516.

Hipel, K.W. and McLeod, A.I. (1994). *Time Series Modelling of Water Resources and Environmental Systems.* Elsevier Science: Amsterdam.

Hipel, K.W., McLeod, A.I., and Lennox, W.C. (1977). Advances in Box-Jenkins modelling, 1, Model construction. *Water Resour. Res.*, **13**, 567–575.

Hipel, K.W., McLeod, A.I., and Li, W.K. (1985). Causal and dynamic relationships between natural phenomena *in Time Series Analysis: Theory and Practice 6*, Eds. O.D. Anderson, J. Kord, and E.A. Robinson. North Holland, Amsterdam.

Hong, Y. (1996a). Testing for independence between two covariance stationary time series. *Biometrika*, **83**, 612–625.

Hong, Y. (1996b). Consistent testing for serial correlation of unknown form. *Econometrica*, **64**, 837–864.

Hong, Y. (1997). One-sided testing for conditional heteroscedasticity in time series models. *J. Time Ser. Anal.*, **18**, 253–277.

Hong, Y. and Shehadeh, R.D. (1999). A new test for ARCH effects and its finite-sample performance. *J. Bus. Econ. Stat.*, **17**, 91–108.

Horváth, L. and Kokoszka, P. (2001). Large sample distribution of weighted sums of ARCH(p) squared residual correlations. *Economet. Theor.*, **17**, 283–295.

Hosking, J.R.M. (1978). A unifed derivation of the asymptotic distribution of goodness-of-fit statistics for autoregressive time-series models. *J. Roy. Statist. Soc.* B, **40**, 341–349.

Hosking, J.R.M. (1980a). Lagrange-multiplier tests of time series models. *J. Roy. Statist. Soc.* B, **42**, 170–181.

Hosking, J.R.M. (1980b). The multivariate portmanteau statistic. *J. Amer. Statist. Ass.*, **75**, 602–607.

Hosking, J.R.M. (1981). Fractional differencing. *Biometrika*, **68**, 165–176.

Hosking, J.R.M. (1981a). Lagrange-multiplier tests of multivariate time-series model. *J. Roy. Statist. Soc.* B, **43**.

Hosking, J.R.M. (1981b). Equivalent forms of the multivariate portmanteau statistic. *J. Roy. Statist. Soc.* B, **43**, 261–262.

Hosking, J.R.M. (1982). Some models of persistence in time series in *Time Series Analysis: Theory and Practice 1.* Ed. O.D. Anderson, pp.641–653, North Holland, Amsterdam.

Hosking, J.R.M. (1984). Modelling persistence in hydrological time series using fractional differencing. *Water Resour. Res.*, **20**, 1898-1908.

Hosking, J.R.M. (1996). Asymptotic distributions of the sample mean, autocovariances, and autocorrelations of long-memory time series. *J. Economet.*, **73**, 261–284.

Hwang, S.Y., Basawa, I.V., and Reeves, J. (1994). The asymptotic distributions of residual autoconditions and related tests of fit for a class of nonlinear time series models. *Statistica Sinica*, **4**, 107–125.

Jacobs, P.A. and Lewis, P.A.W. (1978a). Discrete time series generated by mixture I: correlational and runs properties. *J. Roy. Statist. Soc.* B, **40**, 94–105.

Jacobs, P.A. and Lewis, P.A.W. (1978b). Discrete time series generated by mixture II: asymptotic properties. *J. Roy. Statist. Soc.* B, **40**, 222-228.

Jenkins, G.M. and Watts, D.G. (1968). *Spectral Analysis and Its Applications.* Holden-Day, San Francisco.

Jung, R.C. and Tremayne, A.R. (2003). Testing for serial dependence in time series models of counts. *J. Time Ser. Anal.*, **24**, 65–84.

Kaufmann, H. (1987). Regression models for nonstationary categorical time series: asymptotic estimation theory. *Ann. Statist.*, **15**, 79–98.

Keenan, D.M. (1985). A Tukey non-additivity-type test for time series nonlinearity. *Biometrika*, **72**, 39–44.

Kheoh, T.S. and McLeod, A.I. (1992). Comparison of two modified portmanteau tests for model adequacy. *Comput. Statist. Data Anal.*, **14**, 99–106.

King, M.L. and Wu, P.X. (1990). Locally optimal one-sided tests for multiparameter hypotheses, presented at the 6th World Congress of the Econometric Society, Barcelona; Working Paper 2/90, Monash University, Department of Econometrics.

Klimko, L.A. and Nelson, P.I. (1978). On conditional least squares estimation for stochastic processes. *Ann. Statist.*, **6**, 629–642.

Koenker, R. (1981). A note on studentizing a test for heteroscedasticity. *J. Economet.*, **17**, 107–112.

Koul, H.L. and Stute W. (1999). Nonparametric model checks for time series. *Ann. Statist.*, **27**, 204–236.

Kraft, D.F. and Engle, R.F. (1983). Autogressive conditional heteroskedasticity in multiple time series. Unpublished manuscript, Department of Economics, University of California at San Diego.

Kwan, A.C.C. (1993). A note on the finite-sample distribution of Lagrange multiplier tests for univariate time series models. *Commun. Statist. Simula.*, **22(4)**, 1135–1160.

Kwan, A.C.C. and Wu, Y. (1997). Further results on the finite sample distribution of Monti's portmanteau test for the adequacy of an ARMA(p, q) model. *Biometrika*, **84**, 733–736.

Kwan, A.C.C. and Wu, Y. (2003). A re-examination of the finite sample properties of Peña and Rodriǵuez portmanteau test of lack of fit in time series. Working paper, No. 157, Department of Economics, The Chinese University of Hong Kong, Hong Kong.

LaMarche, V.C., Jr. (1974). Paleoclimatic inferences from long tree ring records. *Science*, **183**, 1042–1048.

Lawrance, A.J. and Kottegoda, N.T. (1977). Stochastic modelling of river flow time series (with discussion). *J. Roy. Statist. Soc.* A, **140**, 1–47.

Lawrance, A.J. and Lewis, P.A.W. (1977). An exponential moving-average sequence and point process (EMA1), *J. Appl. Prob.*, **14**, 98–113.

Lawrance, A.J. and Lewis, P.A.W. (1985). Modelling and residual analysis of nonlinear autoregressive time series in exponential variables (with discussion). *J. Roy. Statist. Soc.* B, **47**, 165–202.

Lawrance, A.J. and Lewis, P.A.W. (1987). Higher-order residual analysis for nonlinear time series with autoregressive correlation structures. *International Statistical Review*, **55**, 21–35.

Ledolter, J. (1983). Comments on *"Model checking in time series analysis"* by Paul Newbold in Applied Time Series Analysis of Economic Data. Ed. A. Zellner, pp144–149. Bureau of the Census, Washington D.C.

Lee, C.H., and Martin, R.D. (1986). Ordinary and proper location M-estimates for autoregressive-moving average models. *Biometrika*, **73**, 679–686.

Lee, J.H.H. (1991). A Lagrange multiplier test for GARCH models. *Economics Letters*, **37**, 265–271.

Lee, J.H.H. and King, M.L. (1993). A locally most mean powerful based score test for ARCH and GARCH. *J. Bus. Econ. Stat.*, **11**, 17–27.

Li, W.K. (1981). Unpublished Ph.D. thesis, University of Western Ontario, London, Canada.

Li, W.K. (1984). On the autocorrelation structure and identification time series. *J. Time Ser. Anal.*, **5**, 173–181.

Li, W.K. (1988). A goodness-of-fit test in robust time series modelling. *Biometrika*, **75**, 355–361.

Li, W.K. (1989). Discriminating nonlinear time series models using a bootstrapped Cox test. *Technical Report*, University of Hong Kong.

Li, W.K. (1991). Testing model adequacy for some Markov regression models for time series. *Biometrika*, **78**, 83–89.

Li, W.K. (1992). On the asymptotic standard errors of residual autocorrelations in nonlinear time series modeling. *Biometrika*, **79**, 435–437.

Li, W.K. (1993). A simple one degree of freedom test for non-linear time series model discrimination. *Statistica Sinica*, **3**, 245–254.

Li, W.K. (1994). Time series models based on generalized linear models: some further results. *Biometrics*, **50**, 506–511.

Li, W.K. and Hui, Y.V. (1988). An algorithm for the exact likelihood of periodic autoregressive moving-average models. *Commun. Statist. Simula.* B, **17**, 1483–1494.

Li, W.K. and Hui, Y.V. (1989). Robust multiple time series modelling. *Biometrika*, **76**, 309–315.

Li, W.K. and Hui, Y.V. (1994). Robust residual cross-correlation tests for lagged relations in time series. *J. Statist. Comput. Simul.*, **49**, 103–109. (http://www.tandf.co.uk/journals/titles/00949655.html)

Li, W.K., Ling, S., and McAleer, M. (2002). Recent theoretical results for time series models with GARCH errors. *J. Econ. Surveys*, **16**, 245–269.

Li, W.K. and Mak, T.K. (1994). On the squared residual autocorrelations in non-linear time series with conditional heteroskedasticity. *J. Time Ser. Anal.*, **15**, 627–636.

Li, W.K. and McLeod, A.I. (1981). Distribution of the residual autocorrelations in multivariate ARMA time series models. *J. Roy. Statist. Soc.* B, **43**, 231–239.

Li, W.K. and McLeod, A.I. (1986). Fractional time series modelling. *Biometrika*, **73**, 217–221.

Li, W.K. and McLeod, A.I. (1988). ARMA modellign with non-Gaussian innovations. *J. Time Ser. Anal.*, **9**, 155–168.

Li, W.K. and Tong, H. (2001). Time Series: Advanced Methods in *International Encyclopedia of Social and Behavioral Sciences.* V23, 15699–15704, Elsevier Science, Oxford.

Li, W.K. and Yu, P. (2003). On the residual autocorrelation of the autoregressive conditional duration model. *Economics Letters*, **79**, 169–175.

Ling, S. and Li, W.K. (1997a). On fractionally integrated autoregressive moving-average time series models with conditional heteroscedasticity. *J. Amer. Statist. Ass.*, **92**, 1184–1193.

Ling, S. and Li, W.K. (1997b). Diagnostic checking of nonlinear multivariate time series with multivariate ARCH errors. *J. Time Ser. Anal.*, **18**, 447–464.

Liu, J. (1992). On stationarity and asymptotic inference of bilinear time series models. *Statistica Sinica*, **2**, 479–494.

Liu, J. and Brockwell, P.J. (1988). On the general bilinear time-series model. *J. Appl. Prob.*, **25**, 553–564.

Ljung, G.M. (1986). Diagnostic testing of univariate time series models. *Biometrika*, **73**, 725–730.

Ljung, G.M. and Box, G.E.P. (1978). On a measure of lack of fit in time series models. *Biometrika*, **65**, 297–303.

Lo, C.L. and Li, W.K. (1990). Two new approaches to robust estimation in time series. *J. Statist. Comput. Simul.*, **34**, 209–255.

Lobato, I.N. (2003). Testing for nonlinear autoregression. *J. Bus. Econ. Stat.*, **21**, 164–173.

Loeve, M. (1978). *Probability Theory* II, 4th edition, Springer-Verlag, New York.

Luukkonen, R., Saikkonen, P., and Teräsvirta, T. (1988). Testing linearity against smooth transition autoregressive models. *Biometrika*, **75**, 491–499.

Mak, T.K., Wong, H., and Li, W.K. (1997). Estimation of nonlinear time series with conditional heteroscedastic variances by iteratively weighted least squares. *Comput. Statist. Data Anal.*, **24**, 169–178.

Mandelbrot, B.B. (1977). *Fractals; form, chance and dimension.* Freeman, San Francisco.

Maronna, R.A. (1976). Robust M-estimators of multivariate location and scatter. *Ann. Statist.*, **4**, 51–67.

Marsaglia, G. (1976). Random number generation. In *Encyclopedia of Computer Science*, Ed. A. Ralston, pp.1192–1197. Van Nostrand Reinhold, New York.

Marshall, R.J. (1980). Autocorrelation estimation of time series with randomly missing observations, *Biometrika*, **67**, 567–570.

Martin, R.D. (1980). Robust estimation of autoregressive models (with discussion). In *Directions in Time Series*, Eds. D.R. Brillinger and G.C. Tiao, pp.228–262. I.M.S., Hayward.

Martin, R.D. (1982). The Cramér-Rao bound and robust M-estimates for autoregression. *Biometrika*, **69**, 437–442.

Martin, R.D., Samarov, A., and Vandaele, W. (1983). Robust methods for ARMA models in *Applied Time Series Analysis of Economic Data*, Ed. A. Zellner, pp.153–169. Bureau of the Census: Washington, D.C.

Masarotto, G. (1987). Robust and consistent estimates of autoregressive-moving average parameters, *Biometrika*, **74**, 791–797.

McAleer, M., McKenzie, C.R., and Hall, A.D. (1988). Testing separate time series models. *J. Time Ser. Anal.*, **9**, 169–189.

McCullagh, P. and Nelder, J.A. (1989). *Generalized Linear Models*, 2nd edition. Chapman and Hall, London.

McKenzie, E. (1985). Some simple models for discrete variate time series. *Water Resour. Bullet.*, **21**, 645–650.

McLeod, A.I. (1975). Derivation of the theoretical autocovariance function of autoregressive-moving average time series. *Appl. Statist.*, **24**, 255-256.

McLeod, A.I. (1977). Improved Box-Jenkins estimators. *Biometrika*, **64**, 531–534.

McLeod, A.I. (1978). On the distribution of residual autocorrelations in Box-Jenkins model. *J. Roy. Statist. Soc.* B, **40**, 296–302.

McLeod, A.I. (1979). Distribution of the residual cross-correlation in univariate ARMA time series models. *J. Amer. Statist. Ass.*, **74**, 849–855.

McLeod, A.I. (1994). Diagnostic checking of periodic autoregression models with application. *J. Time Ser. Anal.*, **15**, 221–233.

McLeod, A.I. and Hipel, K. (1978). Preservation of the rescaled adjusted range, Part 1, A reassessment of the Hurst phenomenon. *Water Resour. Res.*, **14**, 491–508.

McLeod, A.I., Hipel, K.W., and Lennox, W.C. (1977). Advances in Box-Jenkins modelling, 2, applications. *Water Resour. Res.*, **13**, 577–586.

McLeod, A.I. and Li, W.K. (1983). Prewhitening and Power. *Technical Report*, Department of Statistics, University of Hong Kong.

McLeod, A.I. and P.R. Holanda Sales (1983). An algorithm for approximate likelihood calculations of ARMA and Seasonal ARMA models. *Appl. Statist.*, **32**, 211–223.

McQuarrie, A.D.R. and Tsai, C.L. (1998). *Regression and Time Series Model Selection.* World Scientific: Singapore.

Milhoj, A. (1981). A test of fit in time series models. *Biometrika*, **68**, 171–187.

Moeanaddin, R. and Tong, H. (1988). A comparison of likelihood ratio test and CUSUM test for threshold autoregression. *The Statistician*, **37**, 213–225. (Corrigenda and Addenda in *The Statistician*, **37**, 473–474.)

Monti, A.C. (1994). A proposal for residual autocorrelation test in linear models. *Biometrika*, **81**, 776–780.

Moore, D.S. (1978). Chi-square Tests in *Studies in Statistics.* Ed. R.V. Hogg, pp.66–106. The Mathematical Association of America.

Moran, P.A.P. (1948). Some theorems on time series. II. The significance of the serial correlation coefficient. *Biometrika*, **35**, 255–260.

Nelson, D.B. and Cao, C.Q. (1991). Inequality constraints in the univariate GARCH model. *J. Bus. Econ. Stat.*, **10**, 229–235.

Nelson, H.L. and Granger, C.W.J. (1979). Experience with using the Box-Cox transformation when forecasting economic time series. *J. Economet.*, **10**, 57–69.

Newbold, P. (1980). The equivalence of two tests of time series model adequacy. *Biometrika*, **67**, 463–465.

Newton, H.J. (1982). Using periodic autoregressions for multiple spectral estimation. *Technometrics*, **24**, 109–116.

Nicholls, D.F. (1979). Time series analysis of animal populations. In *Interactive Statistics,* Ed. D.R. McNeil, pp.235–246. North Holland, Amsterdam.

Nicholls, D.F. and Quinn, B.G. (1982). *Random Coefficient Autoregressive Models: An Introduction.* Lecture Notes in Statist. **11**, Springer-Verlag, New York.

O'Connell, P.E. and Jones, D.A. (1979). Some experience with the development of models for the stochastic simulation of daily flows. In *Inputs for risk analysis in water systems*, pp.281–314. Water Resour. Publications, Colorado.

Ozaki, T. (1980). Non-linear threshold autoregressive models for non-linear random vibrations. *J. Appl. Prob.*, **17**, 84–93.

Pagano, M. (1978). On periodic and multiple autoregressions. *Ann. Statist.*, **6**, 1310–1317.

Peña, D., Tiao, G.C., and Tsay, R.S. (2001). *A Course in Time Series Analysis.* Wiley Interscience, New York.

Peña, D. and Rodriǵuez, J. (2002). A powerful portmanteau test of lack of fit for time series. *J. Amer. Statist. Ass.*, **97**, 601–610.

Petruccelli, J.D. (1988). A comparison of tests for SETAR-type nonlinearity in time series. *Technical Report.* Worcester Polytechnic Institute.

Petruccelli, J.D. and Davies, N. (1986). A portmanteau test for self exciting threshold autoregressive-type nonlinearity in time series. *Biometrika*, **73**, 687–694.

Pham, T.D. and Tran, L.T. (1981). On the first-order bilinear time series model. *J. Appl. Prob.*, **18**, 617–627.

Pierce, D.A. (1972). Residual correlations and diagnostic checking in dynamic disturbance time series models. *J. Amer. Statist. Ass.*, **67**, 636–640.

Pierce, D.A. (1977). Relationships – and the lack thereof – between economic time series, with special reference to money and interest rates. *J. Amer. Statist. Ass.*, **72**, 11–21.

Pierce, Donald A. (1982). The asymptotic effect of substituting estimators for parameters in certain types of statistics. *Ann. Statist.*, **10**, 475–478.

Pierce, Donald A. (1985). Testing normality in autoregressive models. *Biometrika*, **72**, 293–297.

Pierce, D.A. and Haugh, L.D. (1977). Causality in temporal systems. *J. Economet.*, **5**, 265–293.

Pierce, D.A. and Haugh, L.D. (1979). The characterization of instantaneous causality, a comment. *J. Economet.*, **10**, 257–259.

Pöskitt, D.S. and Tremayne, A.R. (1980). Testing the specification of a fitted autoregressive-moving average model. *Biometrika*, **67**, 359–363.

Pöskitt, D.S. and Tremayne, A.R. (1982). Diagnostics test for multiple time series models. *Ann. Statist.*, **10**, 114–120.

Pötscher, B.M. (1983). Order determination in ARMA-models by Lagrangian multiplier tests. *Ann. Statist.*, **11**, 872–885.

Pregibon, D. (1982). Score tests in GLIM. In GLIM *82: Proceedings of the International Conference on Generalized Linear Models*, Lecture Notes in Statistics 14, Ed. R. Gilchrist, pp. 87–97. Springer, New York.

Price, J.M. (1979). The characterization of instantaneous causality, a correction. *J. Economet.*, **10**, 253–256.

Priestley, M.B. (1980). State-dependent models: a general approach to non-linear time series analysis. *J. Time Ser. Anal.*, **1**, 47–71.

Priestley, M.B. (1988). *Non-linear and Non-Stationary Time Series Analysis.* Academic Press, New York.

Prothero, D.L. and Wallis, K.F. (1976). Modelling macroeconomic time series. *J. Roy. Statist. Soc.* A, **139**, 468–500.

Quenouille, M.H. (1947). A large-sample test for the goodness of fit of autoregressive schemes. *J. Roy. Statist. Soc.* A, **110**, 123–129.

Quenouille, M.H. (1949). Approximate tests of correlation in time series. *J. Roy. Statist. Soc.* B, **11**, 68–84.

Quimpo, R.G. (1967). Stochastic Model for Daily River Flow Sequences. Hydrology papers, Colorado State University.

Rao, C.R. (1973). *Linear Statistical Inference and Its Applications.* Wiley, New York.

Resnick, S. and van Den Berg, E. (2001). A test for non-linearity of time series with infinite variance. *Extremes*, **3**, 145–172.

Robinson, P.M. (1991). Testing for strong serial correlation and dynamic conditional heteroskedasticity in multiple regression. *J. Economet.*, **47**, 67–84.

Rousseewu, P.J. and Leroy, A.M. (1987). *Robust Regression and Oultier Detection.* Wiley, New York.

Runde, R. (1997). The asymptotic null distribution of the Box-Pierce Q-statistic for random variables with infinite variance: an application to German stock returns. *J. Economet.*, **78**, 205–216.

Saikkonen, P. and Luukkonen, R. (1988). Lagrange multiplier tests for testing nonlinearities in time series models. *Scand. J. Statist.*, **15**, 55-68.

Samarov, A. and Taqqu, M.S. (1988). On the efficiency of the sample mean in long memory noise. *J. Time Ser. Anal.*, **9**, 191–200.

Schweppe, F.C. (1965). Evaluation of likelihood functions for Gaussian signals. *I.E.E.E., Trans. Info. Theory*, **11**, 61–70.

Self, S.G. and Liang, K.Y. (1987). Asymptotic properties of maximum likelihood estimators and likelihood ratio tests under nonstandard conditions. *J. Amer. Statist. Ass.*, **82**, 605–610.

Serfling, R.J. (1980). *Approximation Theorems in Statistics.* Wiley, New York.

Shin, D.W. and Lee, J.H. (1996). Distribution of residual autocorrelation in nonstationary autoregressive models. *J. Time Ser. Anal.*, **17**, 105–109.

Silvey, S.D. (1959). The Lagrangian multiplier test. *Ann. Math. Statist.*, **30**, 387–407.

Sims, C.A. (1977). Comment on the "Relationships – and lack thereof – between economic time series", by D.A. Pierce, *J. Amer. Statist. Ass.*, **72**, 23–24.

Smith, R.L. (1986). Maximum likelihood estimation for the NEAR(2) model. *J. Roy. Statist. Soc.* B, **48**, 251–257.

Sowell, F. (1992). Maximum likelihood estimation of stationary univariate fractionally integrated time series models. *J. Economet.*, **53**, 165–188.

Subba Rao., T. (1977). On the estimation of bilinear time series models. *Bull. Inst. Internat. Statist.*, **41**, (Paper presented at the 41st Session of ISI, New Delhi, India).

Subba Rao, T. (1981). On the theory of bilinear time series models. *J. Roy. Statist. Soc.* B, **43**, 244–255.

Subba Rao, T. and Gabr, M.M. (1980). A test for linearity of stationary time series. *J. Time Ser. Anal.*, **1**, 145–158.

Suppes, P. (1979). *A Probabilistic Theory of Causality.* North Holland, Amsterdam.

Terdik, G. (1999). *Bilinear Stochastic Models and Related Problems of Nonlinear Time Series Analysis: A Frequency Domain Approach.* Springer Lecture Notes in Statistics, **142**. Springer-Verlag, New York.

Tong, H. (1978). On a threshold model. In *Pattern Recognition and Signal Processing*, Ed. C.H. Chen, 575–586. Sijthoff and Noordhoff, Amsterdam.

Tong, H. (1983). *Threshold Models in Nonlinear Time Series Analysis.* Springer-Verlag, New York.

Tong, H. (1990). *Nonlinear Time Series: A Dynamical System Approach.* Oxford University Press, Oxford.

Tong, H. and Lim, K.S. (1980). Threshold autoregression, limit cycles and cyclical data (with discussion), 575–586. *J. Roy. Statist. Soc.* B, **42**, 245–292.

Tong, H. and Zhang, Z. (2003). *On time-reversibility of multivariate linear processes.* Technical Report, No. 355. The Department of Statistics and Actuarial Science, University of Hong Kong.

Troutman, B.M. (1979). Some results in periodic autoregression. *Biometrika*, **66**, 219–228.

Tsay, R.S. (1986). Nonlinearity tests for time series. *Biometrika*, **73**, 461–466.

Tsay, R.S. (1988). Outliers, level shifts, and variance changes in time series. *J. Forecasting*, **7**, 1–20.

Tsay, R.S. (1989). Testing and modeling threshold autoregressive processes. *J. Amer. Statist. Ass.*, **84**, 231–240.

Tsay, R.S. (2002). *Analysis of Financial Time Series.* Wiley, New York.

Tse, Y.K. (2002). Residual-based diagnostic for conditional heteroscedasticity model. *Economet. J.*, **5**, 358–373.

Tse, Y.K. and Tsui, A.K.C. (1999). A note on diagnosing multivariate conditional heteroscedasticity models. *J. Time Ser. Anal.*, **20**, 679–691.

Tse, Y.K. and Zuo, X.L. (1997). Testing for conditional heteroscedasticity: some Monte Carlo results. *J. Statist. Comput. Simul.*, **58**, 237–253.

Tukey, J. (1961). Curves as parameters and touch estimation. *Proc. 4th Berkeley Symp. Math. Statist. Probab.*, **1**, 681–684.

Vecchia, A.V. (1985). Maximum likelihood estimation for periodic autoregressive-moving average models. *Technometrics*, **27**, 375–384.

Volterra, V. (1959). *Theory of Functionals and of Integro-Differential Equations.* Dover, New York.

Vuong, Q.H. (1989). Likelihood ratio tests for model selection and nonnested hypotheses. *Econometrika*, **57**, 307–333.

Wahrendorf, J., Becher, H., and Brown, C.C. (1987). Bootstrap comparison of non-nested generalized linear models: applications in survival analysis and epidemiology. *Appl. Statist.*, **36**, 72–81.

Walker, A.M. (1950). Note on a generalization of the large sample goodness of fit test for linear autoregressive schemes. *J. Roy. Statist. Soc.* B, **12**, 102–107.

Walker, A.M. (1952). Some properties of the asymptotic power functions of goodness-of-fit tests for linear autoregressive schemes. *J. Roy. Statist. Soc.* B, **14**, 117–134.

Wei, W.W.S. (1990). *Time Series Analysis: Univariate and Multivariate Methods.* Addison Wesley, New York.

Weiss, A.A. (1986). Asymptotic theory for ARCH models: estimation and testing. *Economet. Theor.*, **2**, 107–131.

Weiss, G. (1975). Time-reversibility of linear stochastic processes. *J. Appl. Prob.*, **12**, 831–836.

White, H. (1994). *Estimation, Inference, and Specification Analysis.* Cambridge University Press, New York.

Whittle, P. (1962). Gaussian estimation in stationary time series. *Bull. Inst. Statist.*, **39**, 105–129.

Wiener, N. (1956). The Theory of Prediction, in *Modern Mathematics for Engineers*, Series 1, Ed. E.F. Beckenback, Ch.8, McGraw-Hill, New York.

Williams, D.A. (1970). Discrimination between regression models to determine the pattern of enzyme synthesis in synchronous cell cultures. *Biometrics*, **28**, 23–32.

Wilson, G.T. (1973). The estimation of parameters in multivariate time series models. *J. Roy. Statist. Soc.* B, **35**, 76–85.

Wong, C.S. and Li, W.K. (1997). Testing of threshold autoregression with conditional heteroscedasticity. *Biometrika*, **84**, 407–418.

Wong, C.S. and Li, W.K. (2000). Testing for double threshold autoregressive conditional heteroscedastic model. *Statistica Sinica*, **10**, 173–189.

Wong, C.S. and Li, W.K. (2001). On a mixture autoregressive conditional heteroscedastic model. *J. Amer. Statist. Ass.*, **96**, 982–995.

Wong, H. and Li, W.K. (1995). Portmanteau test for conditional heteroscedasticy, using ranks of squared residuals. *J. Appl. Statist.*, **22**, 121–134.(http://www.tandf.co.uk/journals/routledge/02664763.html)

Wong, H. and Li, W.K. (1996). Distribution of the cross-correlation of squared residuals in ARIMA models. *The Canadian Journal of Statistics*, **24**, 489–502.

Yajima, Y. (1985). On estimation of long-memory time series models. *Australian J. Statist.*, **27**, 303–320.

Zeger, S.L. (1988). A regression model for time series of counts. *Biometrika*, **75**, 621–629.

Zeger, S.L. and Qaqish, B. (1988). Markov regression models for time series: a quasi-likelihood approach. *Biometrics*, **44**, 1019–1031.

Author index

Subject index